Contents

Cl

CaCl + NaOH → white ppt   Al + NaOH → white
CU(II) + " → blue "   ↓
I(II) + " → Brown "   excess → clear
I(III) + " →   Mg + NaOH → white
                                    NH₃ + " → Nothing

How to use this book

What this book contains

The contents of this book match the specification for WJEC AS Level Chemistry. It provides you with information and practice examination questions that will help you to prepare for the examinations at the end of the year.

This book covers all three of the Assessment Objectives required for your WJEC Chemistry course. The main text covers the three Assessment Objectives:

- AO1 Knowledge and Understanding
- AO2 Application of Knowledge and Understanding.
- AO3 Analyse, interpret and evaluate information, ideas and evidence

This book also addresses:

- The mathematics of chemistry, which will represent a minimum of 20% of your assessment, with explanations and worked examples.
- Practical work. The assessment of your practical skills and understanding of experimental chemistry represents a minimum of 15%, and will also be developed by your use of this book. Some practical tasks are integrated into the chapters.

The book content is clearly divided into the units of this course. These are Unit 1 – The Language of Chemistry, Structure of Matter and Simple Reactions, and Unit 2 – Energy Rate and Chemistry of Carbon Compounds.

Each chapter covers one topic. Each topic is divided into a number of sub-topics, which are listed at the start of each chapter, as a list of learning objectives. Following this, there are a number of practice exam questions designed to help you to practise for the examinations and to reinforce what you have learned. Answers to these questions are given at the end of the book.

Marginal features

The margins of each page hold a variety of features to support your learning:

 Key Terms

These are terms that you need know how to define. They are **highlighted in blue** in the body of the text.

Knowledge check

These are short questions for you to check whether you have followed the material in the text as you go along, and allow you to apply the knowledge you have acquired. Answers are given at the back of the book.

▼ Study point

These contain advice that may help to clarify certain points or aspects of each topic, or help you understand and use the knowledge content.

 **Stretch & Challenge**

This may provide material that while it might not be in the main text, or strictly in the specification and the exam, it will still be relevant to it. It may provide new material that is of interest, and helps to broaden your understanding overall.

Exam tip

This feature provides advice based on the examiner's experience, and highlights certain points that often cause problems for students.

 Link From time to time facts and points appear that are relevant to different parts of the specification, so it will broaden your understanding overall if these connections are made.

YOU SHOULD KNOW ›››

Learning objectives are provided for each main sub-topic.

PRACTICAL CHECK

Occasionally a topic covers an experiment or a practical that is a **specified practical task**. This feature appears alongside in the margin to highlight its importance and to give you some extra information and hints on understanding it fully.

 **Extra Help**

These are helpful hints or extra explanations of key points.

WJEC
Chemistry
for AS Level

Peter Blake

Elfed Charles

Kathryn Foster

Illuminate
Publishing

Published in 2015 by Illuminate Publishing Ltd, P.O. Box 1160, Cheltenham, Gloucestershire GL50 9RW

Orders: Please visit www.illuminatepublishing.com or email sales@illuminatepublishing.com

British Library Cataloguing in Publication Data
A catalogue record for this book is available from the British Library

ISBN 978-1-908682-54-3

Printed by: Wyndeham Grange, Southwick

07.15

The publisher's policy is to use papers that are natural, renewable and recyclable products made from wood grown in sustainable forests. The logging and manufacturing processes are expected to conform to the environmental regulations of the country of origin.

Every effort has been made to contact copyright holders of material produced in this book. If notified, the publisher will be pleased to rectify any errors or omissions at the earliest opportunity.

This material has been endorsed by WJEC and offers high quality support for the delivery of WJEC qualifications. While this material has been through a WJEC quality assurance process, all responsibility for the content remains with the publisher.

The questions used in the exam practice sections of the book are informed by the Specimen Assessment Material (SAM) published by WJEC, but have been written by the authors and reflect the opinion of the authors alone and have not been produced by the examination board.

Editor: Geoff Tuttle
Design: Nigel Harriss
Layout and all original artwork: GreenGate Publishing, Tonbridge

Cover image: © Shutterstock

Acknowledgements

The publisher would like to thank Judith Bonello for her help and advice at key stages in the development process.

MATHS

As assessment of your mathematical skills is very important, this feature demonstrates some common uses of mathematics in chemistry. There is nothing difficult here. You are preparing for a chemistry examination, not a maths exam, but it is still important to apply numerical analysis, and these features will help you to do so. Mathematical requirements are given in Appendix B (AS) or Appendix C (A Level), at the end of the specification course content. The level of understanding is equivalent to Level 2, or GCSE.

HOW SCIENCE WORKS

In some cases it helps you to see how chemistry itself has evolved, the interactions between theory and experiments as well as their limitations. Science works by using theories and ideas, knowledge and understanding, IT and ICT and experimental investigations to obtain, analyse, interpret and evaluate data. Also by considering the applications of science, benefits, risks and ethical issues it evaluates how society may use science to inform decision making.

AS Chemistry – a summary of assessment

Assessment in the AS Specification consists of two written papers of 1 hour 30 minutes each, and there is one paper for each of two units that are themselves each 50% of the qualification. There are 80 marks available on each paper.

Unit 1 deals with the Language of Chemistry, the Structure of matter and Simple reactions.

Unit 2 covers Energy, Rate and the Chemistry of Carbon compounds.

Each paper consists of Section A short answer questions (for 10 marks), and Section B structured and extended answer questions. Section B questions are worth 70 marks.

There are no multiple choice questions in these papers.

Assessment objectives (AOs) and weightings

Assessment objectives

Examination questions are written to reflect the assessment objectives described in the specification. You must meet the following assessment objectives in the context of the subject content, which is given in detail in the specification.

AO1 Covers showing knowledge and understanding of all aspects of the subject.

AO2 Covers applying this knowledge and understanding theoretically, practically and qualitatively and quantitatively.

AO3 Covers analysis, interpretation and evaluation of scientific information and evidence, making judgements, reaching conclusions and developing practical design and procedures.

The weightings of these objectives – which are the same for both units 1 and 2 – are as follows:

AO1 – 17.5%

AO2 – 22.5%

AO3 – 10%

Thus the overall weighting across both papers is as follows:

AO1 – 35%

AO2 – 45%

AO3 – 20%

Mathematical skills

These will be assessed throughout the two papers and have a total weighting of at least 20%. The skills tested include:

- Algebraic manipulation
- Use of calculators
- Means
- Significant figures
- Graph plotting and analysis
- Analysis of spectra
- Understanding 2d and 3d structures in molecular shapes

Practical work

This is an important and intrinsic part of the specification and is covered in two ways: first as a part of the written papers in which its weighting is at least 15% and, secondly, through direct practical work in the laboratory that will prepare candidates to deal with the written work. Although there is no mark as such for the laboratory work, the exercises selected by the teacher must be performed satisfactorily, recorded and inspected.

The type of direct practical work to be undertaken is listed in the specifications of both units and includes the use and application of scientific method and practices, graph plotting, data analysis, errors and precision and the use of instruments and equipment.

Experiments will include preparations, titrations, finding enthalpy changes and measuring rates of reaction.

Suggested practical exercises

Wherever possible we have given an example in relevant topics of the possible type of practical exercise that you might encounter, for example:

Unit 1, Topic 1.6

- Gravimetric analysis, for example, by the precipitation of a Group 2 metal carbonate or a metal chloride
- Identification of unknown solutions by qualitative analysis

Unit 1, Topic 1.7

- Preparation of a soluble salt by titration
- Standardisation of an acid solution
- Back titration, e.g., finding the percentage of $CaCO_3$ in limestone
- Double titration, e.g., analysis of $NaOH/NaCO_3$ mixture

Unit 2, Topic 2.1

- Indirect determination of an enthalpy change such as $MgO + CO_2$ forming $MgCO_3$
- Enthalpy change of combustion

Unit 2, Topic 2.2

- Reaction rate by gas collection
- Iodine 'clock' reaction

Unit 2, Topic 2.6

- Nucleophilic substitution, e.g., 1-bromobutane with NaOH

Unit 2, Topic 2.7

- Preparation and separation of an ester

The examinations

As well as being able to recall facts, name structures and describe their functions, you also need to appreciate the underlying principles of the subject and understand associated concepts and ideas. In other words, you need to develop skills so that you can apply what you have learned, perhaps to situations not previously encountered. For example, the inter-conversion of numerical data and graph form; the analysis and evaluation of numerical data or written information; the interpretation of data; and the explanation of experimental results.

You will be expected to answer different styles of question, in each paper, for example:

- **Section A** Short answer questions – these often require a one-word answer or are a simple calculation.
- **Section B** Structured questions may be in several parts usually about a common theme. They become more difficult as you work your way through. Structured questions can be

short, requiring a one-word response, or may include the opportunity for extended writing. The number of lined spaces and the mark allocation at the end of each part question are there to help you. They indicate the length of answer expected. If three marks are allocated then you must give three separate points.

- In each paper there will be ONE longer six-mark question that will be assessed using a banded level of response mark scheme. What is required is a piece of writing that answers the question directly using well-constructed sentences and suitable chemical terminology. Often candidates rush into such questions. You should take time to read it carefully to discover exactly what the examiner requires in the answer, and then construct a plan. This will not only help you organise your thoughts logically but will also give you a checklist to which you can refer when writing your answer. In this way you will be less likely to repeat yourself, wander off the subject or omit important points.

Further notes on the papers

- There will be no multiple choice questions.
- A maximum of 10% will rely on recall only, i.e. no understanding.
- A minimum of 15% will be related to practical work and a minimum of 20% to Level 2 mathematical skills.

Examination questions are worded very carefully to be clear and concise. It is essential not to penalise yourself by reading questions too quickly or too superficially. Take time to think about the precise meaning of each word in the question so that you can construct a concise, relevant and unambiguous response. To access all the available marks it is essential that you follow the instructions accurately. Here are some words that are commonly used in examinations:

- *Complete:* You may be asked to complete a comparison table. This is straightforward and, if you know your work, you may pick up easy marks. For example: Complete the table to show the number of bonding electrons and the molecular shape. Follow the instructions carefully. If you leave a space blank in such a question, your examiner will not assume that this is equivalent to a cross. Similarly, if you put a tick and change your mind, do not put a line through the tick to convert it to a cross. Cross it out and write a cross.
- *Describe* This term may be used where you need to give a step-by-step account of what is taking place.
- *Explain* A question may ask you to describe and also explain. You will not be given a mark for merely describing what happens – a chemical explanation is also needed.
- *Suggest* This action word often occurs at the end of a question. There may not be a definite answer to the question but you are expected to put forward a sensible idea based on your chemical knowledge.
- *Name* You must give no more than a one-word answer. You

do not have to repeat the question or put your answer into a sentence. That would be wasting time.

- *State* Give a brief, concise answer with no explanation.

- *Compare* If you are asked to make a comparison do so. Make an explicit comparison in each sentence, rather than writing separate paragraphs about what you are comparing.

- *Deduce* Use the information provided and your knowledge to answer the question.

- *Calculate* Work out the answer required using the information provided and your mathematical knowledge.

- *Predict* Evaluate the information provided and use your judgement to give an answer.

- *Write or Balance an equation* To write you will need to know the reactants and products, to balance you will need to apply the ideas of valency and the laws of conservation of atoms.

How to maximise your score

We all vary in speed and natural ability but by attacking the challenge of AS Chemistry in the right way the best possible outcome can be achieved. This book has been written by examiners who have had many years of experience of candidates' performance with the aim of pointing students in the right direction. Here are some of the best tips we have gathered over many years of teaching and examining:

1 Give yourself time. Take each topic slowly, clear up any uncertainties then try the exam practice questions. If you need to then make sure you return to each topic after an interval of time to ensure that you still have it mastered. This may take more than one return trip. It is known that the unconscious mind continues to work and sort learned material so you must give it time. Last-minute cramming is of little use.

2 Be careful to understand what the question is really asking for. Candidates sometimes rush ahead down the wrong track and lose both time and marks. Questions on AO1 will ask you to show that you know and/or understand something; those on AO2 will ask you to apply these and AO3 to analyse, interpret and evaluate something.

   The actual lead words in the questions may include, 'state', 'describe', 'draw', 'name' and 'explain' for AO1, 'calculate', for AO2 and 'suggest' and 'analyse data' for AO3.

3 There is no substitute for work and concentration.

Serena Williams practised thousands of serves to become the number one woman tennis player; Johnny Wilkinson kicked hundreds of practice penalties. Practice trains the mind and produces understanding and enjoyment in mastery of the subject.

Unit 1

Overview
The Language of Chemistry, Structure of Matter and Simple Reactions

1.1 Formulae and equations — p10

- Formulae of common compounds including ionic compounds.
- Assigning oxidation numbers to the atoms in compounds and ions.
- Constructing balanced chemical and ionic equations.

1.2 Basic ideas about atoms — p16

- Atoms contain protons, neutrons and electrons.
- Atoms lose or gain electrons to form ions.
- Radiation occurs due to unstable nuclei and there are three types.
- The rate of radioactive decay is measured by half-life.
- Radioactivity can be harmful but can also be useful in many contexts.
- Atoms contain energy levels or shells consisting of a set of subshells that contain orbitals. Electrons occupy these orbitals.
- Ionisation energies give evidence for electronic configuration of elements.
- The hydrogen emission spectrum arises as electrons move from orbits of high energy levels to orbits of lower energy levels.
- The frequency of the convergence limit of spectral lines can be used to give ionisation energy.
- Energy is proportional to frequency, which is inversely proportional to wavelength.
- Energy increases going from the infrared through visible to the ultraviolet region.

1.3 Chemical calculations — p30

- The masses of atoms are very small so relative atomic (molecular) mass is used for elements (compounds).
- The mass spectrometer can be used to determine relative atomic masses.
- The mole is the chemist's unit of amount.
- The number of atoms in one mole of substance is the Avogadro constant.
- Moles can be converted to mass, concentration and volume.
- Mass of products formed in reactions can be calculated from reactant masses and balanced equations.
- The greater the percentage yield and atom economy of a reaction, the more efficient the process.
- Percentage errors in measurements are used to inform the number of significant figures in calculations.

1.4 Bonding

p45

- Molecules are made of atoms bonded together by ionic, covalent or metallic bonds.
- In ionic bonding, electrons are transferred between atoms to form ions that are held together by electrostatic attraction.
- In covalent bonding, each atom usually donates an electron to an electron pair bond but in coordinate bonding one atom provides both electrons.
- Many bonds are intermediate in nature between ionic and covalent and the word polar is used.
- The polarity of the bond depends on the electronegativity difference between the atoms; the electronegativity value being a measure of the electron-attracting power of the atom.
- Intermolecular forces are much weaker than covalent and ionic bonds within the molecule, and are caused by dipole and induced dipole forces between molecules and control physical properties such as boiling temperature.
- Hydrogen bonds are stronger intermolecular forces involving hydrogen and fluorine, oxygen or nitrogen.
- The shapes of simple molecules are governed by the VSEPR principle and the bond angles in linear, trigonal planar, tetrahedral and octahedral molecules and ions must be known.

1.5 Solid structures

p53

- The crystal structures of sodium and caesium chlorides and the reason for the difference between them.
- The structures of diamond and graphite are to be known and compared as should those of ice and the principle of the structure of the iodine crystal.
- The 'electron sea' model of the structure of metals.
- The relationship between the physical properties of solids, such as melting temperature and electrical conductivity, and structure and bonding in giant and simple molecular solids with ionic, covalent and metallic bonding.

1.6 The periodic table

p56

- The arrangement of the periodic table and the electronic structures of the elements in relation to their position in the s, p and d blocks.
- Oxidation and reduction in terms of electron transfer; identification of such changes.
- The general trends in ionisation energy, electronegativity and melting temperature down groups and across periods.
- The general reactions and trends of the elements of Groups 1 and 2 including reactions of the metallic elements, flame colours and reactions of the Group 2 cations with hydroxide, carbonate and sulfate ions.
- The trends in thermal stability of Group 2 carbonates and hydroxides and the trends in solubility of their hydroxides and sulfates.
- The basic nature of the hydroxides of Groups 1 and 2.
- The reaction of the Group 7 elements (halogens) with metals, their trends in volatility, reactivity in terms of relative oxidising power and redox displacement reactions.
- The reaction of halide ions with aqueous Ag^+ followed by ammonia.
- The use of chlorine and fluoride ions in water treatment and related health issues.
- Practical work on salt formation and crystallisation, gravimetric analysis and qualitative analysis of unknown solutions.

1.7 Simple equilibria and acid-base reactions

p62

- Dynamic equilibrium is when the forward and reverse reactions occur at the same rate.
- A system in equilibrium is affected by changing temperature, pressure or concentration.
- Acids are proton donors, so acidity is a measure of $H^+(aq)$ concentration.
- Because $H^+(aq)$ concentrations are very small the pH scale is used to measure acidity.
- Acids neutralise bases and carbonates to form salts.
- Titrations can be used to form soluble salts as well as to determine concentrations of acids or alkalis.

Unit 1

1.1
Formulae and equations

In chemistry, each element has a symbol. A symbol is a letter or two letters which stand for one atom of the element. Formulae are written for compounds. Formulae consist of the symbols of the elements present and the numbers that show the ratio in which the atoms are present. Using symbols and formulae enables you to write equations for chemical reactions.

Atoms are neither created nor destroyed in a chemical reaction, therefore when you write a chemical equation the same number of atoms of each element must be present on each side of the equation. This is done by balancing the equation.

Content

You should be able to demonstrate and apply knowledge and understanding of:

- Formulae of common compounds and common ions and how to write formulae for ionic compounds.
- Oxidation numbers of atoms in a compound or ion.
- How to construct balanced chemical equations, including ionic equations, with appropriate state symbols.

Formulae of compounds and ions

The formula of a compound is a set of symbols and numbers. The symbols say what elements are present and the numbers give the ratio of the numbers of atoms of the different elements in the compound.

The compound carbon dioxide has the formula CO_2. It contains two oxygen atoms for every carbon atom. The formula for sulfuric acid is H_2SO_4. It contains two hydrogen atoms and four oxygen atoms for every sulfur atom. These compounds consist of molecules in which the atoms are bonded covalently. To show two molecules you write $2H_2SO_4$. The 2 in front of the formula multiplies everything after it. Therefore, in $2H_2SO_4$ there are 4H, 2S and 8O atoms, a total of 14 atoms.

For advanced chemistry you will need to know the formulae of a wide range of compounds. The table below gives a list of the formulae of some common compounds. Since many compounds do not consist of molecules but consist of ions and form through ionic bonding, the list contains both ionic and covalent compounds.

Name	Formula	Name	Formula
Water	H_2O	Sodium hydroxide	NaOH
Carbon dioxide	CO_2	Sodium chloride	NaCl
Sulfur dioxide	SO_2	Sodium carbonate	Na_2CO_3
Methane	CH_4	Sodium hydrogencarbonate	$NaHCO_3$
Hydrochloric acid	HCl	Sodium sulfate	Na_2SO_4
Sulfuric acid	H_2SO_4	Copper(II) oxide	CuO
Nitric acid	HNO_3	Copper(II) sulfate	$CuSO_4$
Ethanoic acid	CH_3CO_2H	Calcium hydroxide	$Ca(OH)_2$
Ammonia	NH_3	Calcium carbonate	$CaCO_3$
Ammonium chloride	NH_4Cl	Calcium chloride	$CaCl_2$

The compound calcium chloride is composed of calcium ions, Ca^{2+}, and chloride ions, Cl^-. There are twice as many chloride ions as calcium ions, so the formula is $CaCl_2$. This is not a molecule of calcium chloride but a formula unit of calcium chloride. For an ionic compound the total number of positive charges must equal the total number of negative charges in one formula unit of the compound.

The table below gives the formulae for common ions that you need to learn.

Positive ions		Negative ions	
Name	Formula	Name	Formula
Ammonium	NH_4^+	Bromide	Br^-
Hydrogen	H^+	Chloride	Cl^-
Lithium	Li^+	Fluoride	F^-
Potassium	K^+	Iodide	I^-
Sodium	Na^+	Hydrogencarbonate	HCO_3^-
Silver	Ag^+	Hydroxide	OH^-
Barium	Ba^{2+}	Nitrate	NO_3^-
Calcium	Ca^{2+}	Oxide	O^{2-}
Magnesium	Mg^{2+}	Sulfide	S^{2-}
Copper(II)	Cu^{2+}	Carbonate	CO_3^{2-}
Iron(II)	Fe^{2+}	Sulfate	SO_4^{2-}
Iron(III)	Fe^{3+}	Phosphate	PO_4^{3-}
Aluminium	Al^{3+}		

Note that non-metals change to end in -ide, but if non-metals combine with oxygen to form negative ions, the negative ion starts with the non-metal and ends in -ate.

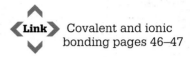 Link Covalent and ionic bonding pages 46–47

YOU SHOULD KNOW ›››
››› the formula of common compounds and ions
››› how to write formula for ionic compounds

 1 Knowledge check
How many atoms of each element are present in:
(a) P_4O_{10}
(b) $2Al(OH)_3$?

 2 Knowledge check
How many oxygen atoms are present in $3Fe(NO_3)_3$?

 3 Knowledge check
Name the following compounds:
(a) Na_2SO_4
(b) $Ca(HCO_3)_2$
(c) $CuCl_2$

The formula for ionic compounds can be calculated by following these steps:

1. Write the symbols of the ions in the compound.

2. Balance the ions so that the total of the positive ions and negative ions adds to zero. (The compound itself must be neutral.)

3. Write the formula without the charges and put the number of ions of each element as a small number following and below the element symbol.

Example 1

Magnesium oxide

1. The ions are Mg^{2+} and O^{2-}.

2. To make the total charge zero, we need one Mg^{2+} ion for every O^{2-} ion ($+2\ -2 = 0$).

3. Formula is MgO (the '1' does not need to be included).

Example 2

Sodium sulfide

1. The ions are Na^+ and S^{2-}.

2. To make the total charge zero, we need two Na^+ ions for every S^{2-} ($+1\ +1\ -2 = 0$) i.e. Na^+ Na^+ S^{2-}.

3. Formula is Na_2S.

Example 3

Calcium nitrate

1. The ions are Ca^{2+} and NO_3^-.

2. Two NO_3^- ions are needed to balance the charge on one Ca^{2+} ion ($-1\ -1\ +2 = 0$) i.e. Ca^{2+} NO_3^- NO_3^-.

3. Formula is $Ca(NO_3)_2$ (note the use of a bracket around the NO_3 before adding the 2).

Oxidation numbers

As you have seen on the previous pages, there are differences in the ratios of the atoms that combine together to form compounds, e.g. H_2O and HCl. A method of expressing the combining power of elements is **oxidation number**. The oxidation number of an element is the number of electrons that need to be added to (or taken away from) an element to make it neutral.

E.g. the iron ion, Fe^{2+}, needs the addition of two electrons to make a neutral atom, therefore it has the oxidation number +2. The chloride ion, Cl^-, needs to lose an electron to make a neutral atom; therefore it has the oxidation number –1.

The use of oxidation numbers can be extended to covalent compounds. Some elements are assigned positive oxidation numbers and others are assigned negative oxidation numbers in accordance with certain rules.

Rule	Example
The oxidation number of an uncombined element is zero.	Metallic copper, Cu: oxidation number 0 Oxygen gas, O_2: oxidation number 0.
The sum of the oxidation numbers in a compound is zero. In an ion the sum equals the overall charge.	In CO_2 the sum of the oxidation numbers of carbon and oxygen is 0. In NO_3^- the sum of the oxidation numbers of nitrogen and oxygen is -1.
In compounds the oxidation numbers of Group 1 metals is $+1$ and Group 2 metals is $+2$.	In $MgBr_2$ the oxidation number of magnesium is $+2$ (oxidation number of each bromine is -1).
The oxidation number of oxygen is -2 in compounds except with fluorine or in peroxides (and superoxides).	In SO_2 the oxidation number of each oxygen is -2 (oxidation number of sulfur is $+4$). In H_2O_2 the oxidation number of oxygen is -1 (oxidation number of hydrogen is $+1$).
The oxidation number of hydrogen is $+1$ in compounds except in metal hydrides.	In HCl the oxidation number of hydrogen is $+1$ (oxidation number of chlorine is -1). In NaH the oxidation number of hydrogen is -1 (oxidation number of sodium is $+1$).
In chemical species with atoms of more than one element, the most electronegative element is given the negative oxidation number.	In CCl_4, chlorine is more electronegative than carbon, so the oxidation number of each chlorine is -1 (oxidation number of carbon is $+4$).

Oxidation numbers are used in redox reactions to show which species is oxidised and which one is reduced.

Oxidation numbers are used to name compounds unambiguously, e.g. potassium, nitrogen and oxygen can combine to give two different compounds, KNO_3 and KNO_2. Since the oxidation number of potassium is $+1$ and that of oxygen is -2, the oxidation number of nitrogen must be $+5$ in KNO_3 and $+3$ in KNO_2. Therefore KNO_3 is called potassium nitrate(V) and KNO_2 is called potassium nitrate(III).

Chemical and ionic equations

Chemical equations are written to sum up what happens in a chemical reaction. When magnesium ribbon is burnt in air there is a bright flame and some white ash is left. In chemical language, this is what happens: magnesium (a solid metal) burns in oxygen (a gas from the air) to form magnesium oxide (a white solid). This can be summed up by writing:

magnesium + oxygen $\longrightarrow$ magnesium oxide

Replacing names with formulae gives a chemical equation for the reaction:

$$Mg + O_2 \longrightarrow MgO$$

(Remember that oxygen exists as a diatomic molecule so its formula is O_2.)

However, since atoms are neither created nor destroyed in a chemical reaction, there must be the same number of atoms of each element on each side of the chemical equation. Counting the number of atoms we see that the equation is not balanced.

Number of atoms on L.H.S. = 1Mg + 2O

Number of atoms on R.H.S. = 1Mg + 1O

How do we balance the equation? It is tempting to write O for the oxygen on the L.H.S. or to write MgO_2 for the magnesium oxide on the R.H.S. You must not do this.

YOU SHOULD KNOW ›››

››› how to assign oxidation numbers to the atoms in compounds or ions

Knowledge check

What is the oxidation number of:

(a) nitrogen in NH_3

(b) phosphorus in P_4

(c) manganese in MnO_4^-

(d) chromium in $K_2Cr_2O_7$?

▼ **Study point**

Sometimes the term oxidation state is used instead of oxidation number. The only difference is that we say for Fe^{2+}, for example, the oxidation number of Fe is $+2$ but the oxidation state of this ion is written as Fe(II).

Link Electronegativity page 48

Link Oxidation and reduction page 58

YOU SHOULD KNOW ›››

››› how to construct balanced chemical and ionic equations

Neither of these are correct formulae for the substances, so they cannot be used. Never change a formula. All you can do to balance an equation is to multiply a formula by putting a number in front of the formula.

To balance the oxygen atoms we need 2 atoms on the R.H.S. so multiply the R.H.S. by 2:

$$Mg + O_2 \longrightarrow 2MgO$$

Number of atoms on L.H.S. = 1Mg + 2O

Number of atoms on R.H.S. = 2Mg + 2O

We need 2 atoms of magnesium on the L.H.S. so multiply the magnesium on the L.H.S. by 2:

$$2Mg + O_2 \longrightarrow 2MgO$$

Number of atoms on L.H.S. = 2Mg + 2O

Number of atoms on R.H.S. = 2Mg + 2O

The equation is balanced.

To give more information state symbols can be included in the equation. The state symbols used are: (s) for solid, (l) for liquid, (g) for gas. A solution in water is described as aqueous, so (aq) is used for a solution.

The final equation becomes:

$$2Mg(s) + O_2(g) \longrightarrow 2MgO(s)$$

The steps in writing a balanced chemical equation are:

1. Write a word equation for the reaction (optional).
2. Write the symbols and formulae for the reactants and products (make sure that all formulae are correct).
3. Balance the equation by multiplying formulae if necessary (never change a formula).
4. Check your answer.
5. Add state symbols (if required).

Exam tip

Remember that all the elements in the mnemonic HOFBrINCl exist as diatomic molecules.

6

Knowledge check

Balance the following equations
(a) $SO_2 + O_2 \longrightarrow SO_3$
(b) $Fe_2O_3 + CO \longrightarrow Fe + CO_2$

Example

Sodium carbonate reacts with dilute hydrochloric acid to give carbon dioxide and a solution of sodium chloride.

Write a balanced chemical equation including state symbols for this reaction.

$$\frac{sodium}{carbonate} + \frac{hydrochloric}{acid} \longrightarrow \frac{sodium}{chloride} + \frac{carbon}{dioxide} + water$$

Writing the formulae gives:

$$Na_2CO_3 + HCl \longrightarrow NaCl + CO_2 + H_2O$$

Number of atoms on L.H.S. = 2Na + 1C + 3O + 1H + 1Cl

Number of atoms on R.H.S. = 1Na + 1C + 3O + 2H + 1Cl

Start by balancing Na atoms, so multiply NaCl on the R.H.S. by 2:

$$Na_2CO_3 + HCl \longrightarrow 2NaCl + CO_2 + H_2O$$

Number of atoms on L.H.S. = 2Na + 1C + 3O + 1H + 1Cl

Number of atoms on R.H.S. = 2Na + 1C + 3O + 2H + 2Cl

Next balance H atoms by multiplying HCl on the L.H.S. by 2:

$$Na_2CO_3 + 2HCl \longrightarrow 2NaCl + CO_2 + H_2O$$

Number of atoms on L.H.S. = 2Na + 1C + 3O + 2H + 2Cl

Number of atoms on R.H.S. = 2Na + 1C + 3O + 2H + 2Cl

Equation is now balanced. Add state symbols.

$$Na_2CO_3(s) + 2HCl(aq) \longrightarrow 2NaCl(aq) + CO_2(g) + H_2O(l)$$

Ionic equations

Many reactions involve ions in solutions. However, in these reactions not all of the ions take part in any chemical change. An ionic equation may help to show what is happening.

Ionic equations are frequently used for displacement and precipitation reactions.

Example 1

When zinc powder is added to copper(II) sulfate solution, copper is displaced and a red-brown deposit is formed on the zinc.

The chemical equation for the reaction is:

$$Zn(s) + CuSO_4(aq) \longrightarrow ZnSO_4(aq) + Cu(s)$$

Writing out all of the ions gives:

$$Zn(s) + Cu^{2+}(aq) + SO_4^{2-}(aq) \longrightarrow Zn^{2+}(aq) + SO_4^{2-}(aq) + Cu(s)$$

There is repetition here. The $SO_4^{2-}(aq)$ ions have not taken part in any chemical change at all. They have been present unchanged throughout. They are called spectator ions and are left out of the ionic equation, which is written:

$$Zn(s) + Cu^{2+}(aq) \longrightarrow Zn^{2+}(aq) + Cu(s)$$

An ionic equation provides a shorter equation which focuses attention on the changes taking place.

Example 2

When a solution of sodium hydroxide is added to a solution of magnesium chloride a white precipitate forms.

The chemical equation for the reaction is:

$$2NaOH(aq) + MgCl_2(aq) \longrightarrow 2NaCl(aq) + Mg(OH)_2(s)$$

Writing out all of the ions gives:

$$2Na^+(aq) + 2OH^-(aq) + Mg^{2+}(aq) + 2Cl^-(aq) \longrightarrow 2Na^+(aq) + 2Cl^-(aq) + Mg(OH)_2(s)$$

The $Na^+(aq)$ ions and the $Cl^-(aq)$ ions do not change during the reaction. They are spectator ions and can be omitted, giving the ionic equation:

$$Mg^{2+}(aq) + 2OH^-(aq) \longrightarrow Mg(OH)_2(s)$$

Stretch & Challenge

Balance the following equation:

$$Cu + HNO_3 \longrightarrow Cu(NO_3)_2 + NO + H_2O.$$

7

Knowledge check

When a solution of sodium sulfate is added to a solution of barium chloride, a white precipitate of barium sulfate forms. Write an ionic equation, including state symbols, for this reaction.

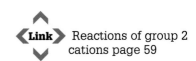

Link Reactions of group 2 cations page 59

Unit 1

1.2
Basic ideas about atoms

Chemistry is the study of how matter behaves. We know that all matter is made up of very small particles called atoms. The idea of atoms was put forward by the Greek philosopher Democritus at the beginning of the fifth century BCE but it was not based on experimental results. It was not until the beginning of the nineteenth century that the atomic theory was revived by John Dalton. By this time, a large number of experimental results had been built up and he was able to provide indirect evidence that matter is made up of atoms. By the late nineteenth and early twentieth century, scientists such as Thomson, Rutherford and Chadwick showed that atoms have an internal structure comprising protons, neutrons and electrons.

Today scientists believe that there are two basic types of particle – quarks and leptons. Protons and neutrons are made from quarks, and electrons belong to the lepton particle family. This unit looks at protons, neutrons and electrons. It shows what happens when an unstable atom splits to form smaller particles and how ionisation energies and emission spectra provide evidence for electronic configuration.

Content

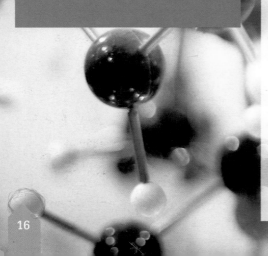

You should be able to demonstrate and apply knowledge and understanding of the:

- Nature of radioactive decay and the resulting changes in atomic number and mass number (including positron emission and electron capture).
- Behaviour of α-, β- and γ-radiation in electric and magnetic fields and their relative penetrating power.
- Half-life of radioactive decay.
- Adverse consequences for living cells of exposure to radiation and use of radioisotopes in many contexts, including health, medicine, radio-dating, industry and analysis.
- Significance of standard molar ionisation energies of gaseous atoms and their variation from one element to another.
- Link between successive ionisation energy values and electronic structure.
- Shapes of s- and p-orbitals and the order of s-, p- and d-orbital occupation for elements 1–36.
- Origin of emission and absorption spectra in terms of electron transitions between atomic energy levels.
- Atomic emission spectrum of the hydrogen atom.
- Relationship between energy and frequency ($E = hf$) and that between frequency and wavelength ($f = c/\lambda$).
- Order of increasing energy of infrared, visible and ultraviolet light.
- Significance of the frequency of the convergence limit of the Lyman series and its relationship with the ionisation energy of the hydrogen atom.

Atomic structure

Atomic structure is not specifically mentioned in the specification. However, since all learners are expected to demonstrate knowledge and understanding of standard content covered at GCSE level, pages 17–18 give a recap on the minimum knowledge that is required about the structure of the atom and how elements and ions are represented.

Atoms are made up of three fundamental particles, the proton, the neutron and the electron.

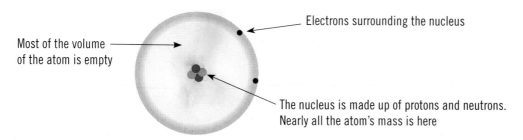

Electrons surrounding the nucleus

Most of the volume of the atom is empty

The nucleus is made up of protons and neutrons. Nearly all the atom's mass is here

The masses and charges of these particles are very small and so are inconvenient, therefore we call the mass of a proton 1, its charge +1 and we describe the other particles relative to these values.

Particle	Relative mass	Relative charge
proton	1	+1
neutron	1	0
electron	Negligible (1/1840)	−1

An atom is electrically neutral because the number of negative electrons surrounding the nucleus equals the number of positive protons in the nucleus.

Representing elements and ions

All atoms of the same element contain the same number of protons. The number of protons in the nucleus of an atom determines the element to which the atom belongs and is known as the **atomic number**.

It is also useful to have a measure for the total number of particles in the nucleus of an atom. This is called the **mass number**.

The full symbol for an element incorporates the atomic number, mass number and symbol

e.g.
$$\text{mass number} \longrightarrow 23$$
$$\mathbf{Na} \longleftarrow \text{symbol}$$
$$\text{atomic number} \longrightarrow 11$$

Atoms of the same element are not all identical. They always have the same number of protons, but they can have different numbers of neutrons. Such atoms are called **isotopes**. Most elements exist naturally as two or more different isotopes. For example, chlorine consists of two isotopes, one having a mass number of 35 and one having a mass number of 37 or $^{35}_{17}\text{Cl}$ and $^{37}_{17}\text{Cl}$.

A particle where the number of electrons does not equal the number of protons is no longer an atom but is called an **ion** and has an electrical charge.

YOU SHOULD KNOW ›››

››› the charges and masses of protons, neutrons and electrons

››› the meaning of atomic number, mass number and isotopes

››› how ions form

Key Terms

Atomic number (Z) is the number of protons in the nucleus of an atom.

Mass number (A) is the number of protons + the number of neutrons in the nucleus of an atom.

Isotopes are atoms having the same number of protons but different numbers of neutrons.

Ion is a particle where the number of electrons does not equal the number of protons.

Exam tip

Don't forget, in any atom:

The atomic number = the number of protons.

The number of protons = the number of electrons.

The number of neutrons = the mass number – the atomic number.

▼ Study point

It is incorrect to state that atomic number = the number of protons and electrons.

Isotopes of an element have the same chemical properties.

8

Knowledge check

Give the number of protons, neutrons and electrons in the two main isotopes of copper: Cu-63 and Cu-65.

Link Ionic bonding
page 47

Knowledge check

State the number of protons and electrons in

(a) $^{131}I^-$

(b) $^{25}Mg^{2+}$.

 Stretch & Challenge

Nuclei contain protons packed together in a very small space. Why do nuclei not fly apart?

! Extra Help

If you are given an element's mass number and symbol, use the periodic table for its atomic number. Remember it might be an isotope so the mass number might be different from that in the periodic table.

 Key Terms

α-particles cluster of 2 protons and 2 neutrons, therefore positively charged.

β-particles fast moving electrons, therefore negatively charged.

γ-rays high energy electromagnetic radiation, therefore no charge.

 **Stretch & Challenge**

β particles can be considered as being formed when a neutron changes into a proton i.e.

$$_0^1n \longrightarrow {}_1^1p + {}_{-1}^0\beta.$$

If a neutral atom loses one or more electrons it forms a positive ion or cation,

e.g. $Na \longrightarrow Na^+ + e^-$

If a neutral atom gains one or more electrons it forms a negative ion or anion,

e.g. $Cl + e^- \longrightarrow Cl^-$

In both examples the number of protons has not changed but the number of electrons has.

The number of electrons in Na^+ is 10 (atomic number − charge on ion).

The number of electrons in Cl^- is 18 (atomic number + charge on ion).

Radioactivity

Types of radioactive emission and their behaviour

Some isotopes are unstable and split up to form smaller atoms. The nucleus divides and sometimes protons, neutrons and electrons fly out. The process is called radioactive decay and the element is said to be radioactive. Radioactive isotopes have unstable nuclei and they give off three types of radiation: **alpha (α)**, **beta (β)** and **gamma (γ)**.

Alpha particles consist of two protons and two neutrons and are therefore helium nuclei. They are the least penetrating of the three and are stopped by a thin sheet of paper or even a few centimetres of air.

Beta particles consist of streams of high-energy electrons and are more penetrating. They can travel through up to 1 m of air but are stopped by a 5 mm thick sheet of aluminium.

Gamma rays are high-energy electromagnetic waves and are the most penetrating of the three radiations and can pass through several centimetres of lead or more than a metre of concrete.

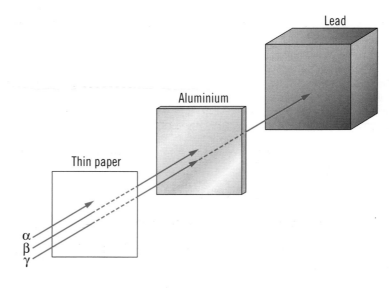

▲ The penetrating powers of radiation

When alpha, beta and gamma radiation pass through matter they tend to knock electrons out of atoms ionising them. Alpha particles are strongly ionising because they are large, relatively slow moving and carry two positive charges. On the other hand, gamma rays are weakly ionising.

Ionisation involves a transfer of energy from the radiation passing through the matter to the matter itself. As the alpha particle is the most strongly ionising of the radiations, this transfer happens most rapidly and so they are the least penetrating. Conversely, since gamma rays are the least ionising they are the most penetrating of the radiations.

When alpha, beta and gamma radiations pass through an electric field, gamma rays are undeflected, while alpha particles are deflected towards the negative charge and beta particles towards the positive charge.

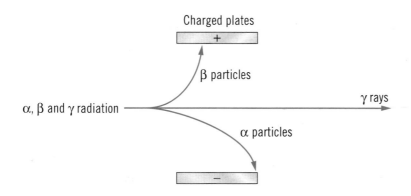

▲ The effect of electric field on radiation

Magnetic fields have a similar effect on alpha, beta and gamma radiation. When a charged particle cuts through a magnetic field it experiences a force referred to as the motor effect.

Alpha particles are deflected by a magnetic field confirming that they must carry a charge. The direction of deflection (which can be determined by Fleming's left-hand rule) demonstrates that they must be positively charged.

Beta particles are deflected by a magnetic field in an opposite direction to alpha particles confirming they must hold a charge opposite to alpha particles.

Gamma rays are unaffected by a magnetic field. This shows gamma rays are uncharged as they do not experience a force when passing through the lines of a magnetic field.

Study point

In equations: $^{4}_{2}He^{2+}$ is acceptable for $^{4}_{2}\alpha$.

$^{0}_{-1}e$ is acceptable for $_{-1}\beta$.

Study point

Electron capture can be regarded as an equivalent to positron emission, since capture of an electron results in the same transmutation as emission of a positron.

Effect on mass number and atomic number

α and β particle emissions result in the formation of a new nucleus with a new atomic number therefore the product is a different element.

When an element emits an α particle its mass number decreases by 4 and its atomic number decreases by 2.

$$^{238}_{92}U \longrightarrow {}^{234}_{90}Th + {}^{4}_{2}\alpha$$

The product is two places to the left in the periodic table.

When an element emits a β particle its mass number is unchanged and its atomic number increases by 1.

$$^{14}_{6}C \longrightarrow {}^{14}_{7}N + {}^{0}_{-1}\beta$$

The product is one place to the right in the periodic table.

A process of inverse beta decay can also occur. This is known as **electron capture**. In the process of electron capture, one of the orbital electrons is captured by a proton in the nucleus, forming a neutron and emitting an electron neutrino (ν_e).

Stretch & Challenge

Because positron emission decreases proton number relative to neutron number, positron decay happens typically in large 'proton-rich' radionuclides. Electron capture is an alternative decay mode for radioactive isotopes with insufficient energy to decay by positron emission. It therefore occurs much more often in smaller atoms than positron emission. Electron capture always competes with positron emission, however it occurs as the only type of beta decay in proton-rich nuclei when there is not enough decay energy to support positron emission.

10

Give the mass number and symbol of the isotope formed when ^{234}Th decays by β emission.

$$^{40}_{19}K + e^- \longrightarrow {}^{40}_{18}Ar$$

The product is one place to the left in the periodic table.

Positron emission or **β⁺ decay** is another subtype of beta decay. In this process a proton is converted into a neutron while releasing a positron and an electron neutrino. The positron is a type of beta particle (β⁺).

$$^{23}_{12}Mg \longrightarrow {}^{23}_{11}Na + β^+$$

The product is one place to the left in the periodic table.

YOU SHOULD KNOW ›››

››› the nature of alpha (α), beta (β) and gamma (γ) radiation

››› their behaviour in electric fields and their relative penetrating power

››› how α- and β- emission affect atomic number and mass number

››› what is meant by half-life

Half-life

The rate at which a radioactive isotope decays cannot be speeded up or slowed down, it is proportional to the number of radioactive atoms present. The nature of radioactive decay is shown below.

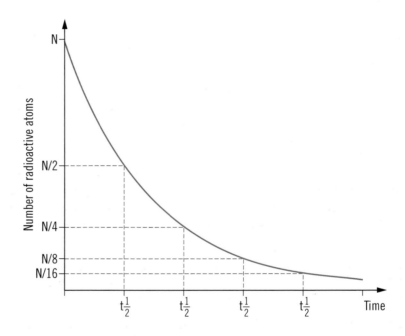

Key Term

Half-life is the time taken for half the atoms in a radioisotope to decay or the time taken for the radioactivity of a radioisotope to fall to half its initial value.

▲ Radioactive decay

The time taken for *N* atoms to decay to *N*/2 atoms is the same as the time taken for *N*/2 atoms to decay to *N*/4 atoms and for *N*/4 atoms to decay to *N*/8 atoms. The time taken to decay to half the number of radioactive atoms is known as **half-life**.

The process resembles a knock out competition such as Wimbledon where one half of the competitors (atoms) disappears over each round (half-life). The number of competitors disappearing during each round (number of atoms decaying each half-life) gets smaller and smaller but is always one half of those remaining.

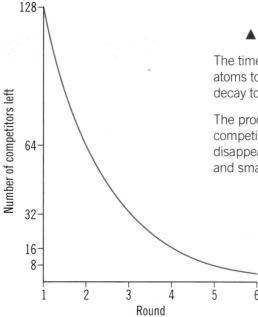

There are two types of calculation involving half-life:

- Finding the time taken for the radioactivity of a sample to fall to a certain fraction of its initial value.
- Finding the mass of a radioactive isotope remaining after a certain length of time given the initial mass.

Example

The radioactive isotope ^{28}Mg has a half-life of 21 hours.

(a) Calculate how long it will take for the activity of the isotope to decay to ⅛ its original value.

(b) If you started with 2.0 g of ^{28}Mg, calculate the mass of this isotope remaining after 84 hours.

Answer

(a) $1 \xrightarrow{21} \dfrac{1}{2} \xrightarrow{21} \dfrac{1}{4} \xrightarrow{21} \dfrac{1}{8}$

$21 \times 3 = 63$ hours

(b) 84 hours = 4 half-lives

$2.0g \xrightarrow{21} 1.0g \xrightarrow{21} 0.5g \xrightarrow{21} 0.25g \xrightarrow{21} 0.125g$

▼ Study point

The greater the half-life of a radioactive isotope the greater the concern since the radioactivity of the isotope exists for a longer time.

11

Knowledge check

An isotope of iodine ^{131}I has a half-life of 8 days. Calculate how long it would take for 1.6 g of ^{131}I to be reduced to 0.20 g of ^{131}I.

Consequences for living cells

Radioactive emissions are potentially harmful. However, we all receive some radiation from the normal background radiation that occurs everywhere. Workers in industries where they are exposed to radiation from radioactive isotopes are carefully monitored to ensure that they do not receive more radiation than is allowed under internationally agreed limits.

Ionising radiation may damage the <u>DNA of a cell</u>. Damage to the DNA may lead to changes in the way the cell functions, which can cause mutations and the formation of cancerous cells at lower doses or cell death at higher doses.

Personal danger from ionising radiation may come from sources outside or inside the body. With a source outside the body gamma radiation is likely to be the most hazardous. However, the opposite is true for sources inside the body and if alpha particle emitting isotopes are ingested they are far more dangerous than an equivalent activity of beta emitting or gamma emitting isotopes.

YOU SHOULD KNOW ›››

- ››› how radiation affects living cells

- ››› some beneficial uses of radioactivity

Stretch & Challenge

Why is α radiation the most harmful if ingested but least harmful outside the body?

Beneficial uses of radioactivity

Although radiation from radioisotopes is harmful to health, at the same time many beneficial uses of radioactivity have been found.

Medicine

- Cobalt-60 in radiotherapy for the treatment of cancer. The high energy of γ-radiation is used to kill cancer cells and prevent a malignant tumour from developing.
- Technetium-99m is the most commonly used medical radioisotope. It is used as a tracer, normally to label a molecule which is preferentially taken up by the tissue to be studied.

All living have Same amount of C¹⁴ trust me

Radio-dating

- Carbon-14 (half-life 5570 years) is used to calculate the age of plant and animal remains. All living organisms absorb carbon, which includes a small proportion of the radioactive carbon-14. When an organism dies there is no more absorption of carbon-14 and that which is already present decays. The rate of decay decreases over the years and the activity that remains can be used to calculate the age of organisms.

- Potassium-40 (half-life 1300 million years) is used to estimate the geological age of rocks. Potassium-40 can change into argon-40 by the nucleus gaining an inner electron. Measuring the ratio of potassium-40 to argon-40 in a rock gives an estimate of its age.

Analysis

- Dilution analysis. The use of isotopically labelled substances to find the mass of a substance in a mixture. This is useful when a component of a complex mixture can be isolated from the mixture in the pure state but cannot be extracted quantitatively.

- Measuring the thickness of metal strips or foil. The metal is placed between two rollers to get the right thickness. A radioactive source (a β emitter) is mounted on one side of the metal with a detector on the other. If the amount of radiation reaching the detector increases, the detector operates a mechanism for moving the rollers apart and vice versa.

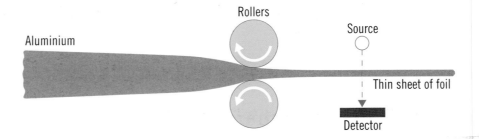

Key Term

An atomic orbital is a region in an atom that can hold up to two electrons with opposite spins.

Exam tip

You need to know the electronic configuration for the first 36 elements.

The configurations for chromium and copper are not as expected, they both end in 4s¹.

The 4s orbitals are filled before the 3d orbitals.

Electronic structure

Electrons hold the key to almost the whole of chemistry since only electrons are involved in the changes that happen during chemical reactions. Electrons within atoms occupy fixed energy levels or shells. Shells are numbered 1, 2, 3, 4, etc. These numbers are known as principal quantum numbers, n. The lower the value of n, the closer the shell to the nucleus and the lower the energy level.

In a shell there are regions of space around the nucleus where there is a high probability of finding an electron of a given energy. These regions are called **atomic orbitals**. Orbitals of the same type are grouped together in a subshell. Each orbital can contain two electrons. Along with charge, electrons have a property called 'spin'. In order for two electrons to exist in the same orbital they must have opposite spins: this reduces the effect of repulsion. Each orbital has its own three-dimensional shape.

There is only one type of s orbital and it is spherical

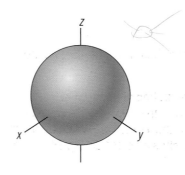

normally drawn as:

There are three different p orbitals (dumbell shaped lobes) known as the p_x, p_y and p_z orbitals. They are at right angles to each other.

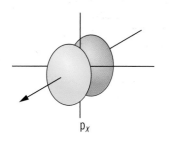 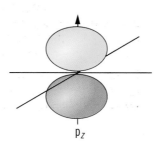 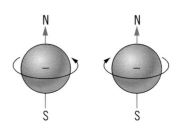

You can visualise an electron spinning like the Earth on its axis.

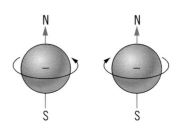

▲ Representation of opposite spins of electrons

These are represented as

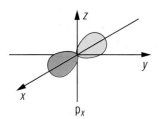 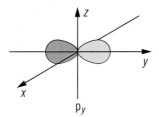 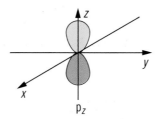

▲ Representation of p orbitals

There are five different d orbitals and seven different f orbitals.

Spin = ↑

Therefore:

- an s subshell can hold 2 electrons
- a p subshell can hold 6 electrons
- a d subshell can hold 10 electrons
- an f subshell can hold 14 electrons.

Filling shells and orbitals with electrons

The way in which an atom's electrons are arranged in its atomic orbitals is called electronic structure or configuration. The electronic structure can be worked out using three basic rules:

1. Electrons fill atomic orbitals in order of increasing energy (Aufbau principle).

2. A maximum of two electrons can occupy any orbital each with opposite spins (Pauli exclusion principle).

3. The orbitals will first fill with one electron each with parallel spins, before a second electron is added with the paired spin (Hund's rule).

The order of filling is shown in the diagram on the right.

An expected order is followed up to the 3p subshell, but then there is a variation, as the 4s subshell is filled before the 3d.

The most common way of representing the **electronic configuration** of an atom is to write the occupied subshells in order of increasing energy with the number of electrons following as a superscript, e.g. nitrogen has two electrons in the 1s orbital, two electrons in the 2s orbital and three electrons in the 2p subshell so it is shown as $1s^2\ 2s^2\ 2p^3$.

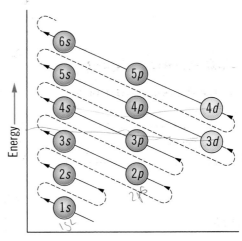

▲ The order of electron filling

▼ Study point

The reason for the 4s orbitals filling before the 3d orbitals is due to increasingly complex influences of nuclear attractions and electron repulsions upon individual electrons.

 **Key Term**

Electronic configuration is the arrangement of electrons in an atom.

Knowledge check

(a) Use electrons in boxes to write the electronic configuration of:

   (i) an atom of phosphorus, P.

   (ii) a magnesium ion, Mg^{2+}.

(b) Write the electronic configuration in terms of sub-shells for a chromium atom.

Knowledge check

State the number of different orbitals in the third quantum shell.

<Link> Position in periodic table page 57

For convenience sake we represent $1s^2\ 2s^2\ 2p^6\ 3s^2\ 3p^6$ as [Ar] rather than write it out each time, e.g. the electronic configuration of manganese, atomic number 25, can be written as $[Ar]\ 3d^5\ 4s^2$.

A convenient way of representing electronic configuration is using 'electrons in boxes'. Each orbital is represented as a box and the electrons as arrows with their clockwise or anticlockwise spins as ↑ or ↓.

Here are the 'electrons in boxes' notation and shorter form of electronic structure for the first ten elements.

Element	Electronic configuration			Electrons in boxes				
				1s	2s	2p		
H	$1s^1$			↑				
He	$1s^2$			↑↓				
Li	$1s^2$	$2s^1$		↑↓	↑			
Be	$1s^2$	$2s^2$		↑↓	↑↓			
B	$1s^2$	$2s^2$	$2p^1$	↑↓	↑↓	↑		
C	$1s^2$	$2s^2$	$2p^2$	↑↓	↑↓	↑	↑	
N	$1s^2$	$2s^2$	$2p^3$	↑↓	↑↓	↑	↑	↑
O	$1s^2$	$2s^2$	$2p^4$	↑↓	↑↓	↑↓	↑	↑
F	$1s^2$	$2s^2$	$2p^5$	↑↓	↑↓	↑↓	↑↓	↑
Ne	$1s^2$	$2s^2$	$2p^6$	↑↓	↑↓	↑↓	↑↓	↑↓

▲ Table of electronic configuration

The electronic configuration of ions is presented in the same way as that of atoms.

Positive ions form by the loss of electrons from the highest energy orbitals so these ions have fewer electrons than the parent atom.

Negative ions form by adding electrons to the highest energy orbitals so these ions have more electrons than the parent atom,

e.g. Na $1s^2\ 2s^2\ 2p^6\ 3s^1$ Na^+ $1s^2\ 2s^2\ 2p^6$

      Cl $1s^2\ 2s^2\ 2p^6\ 3s^2\ 3p^5$ Cl^- $1s^2\ 2s^2\ 2p^6\ 3s^2\ 3p^6$

Ionisation energies

YOU SHOULD KNOW ›››

››› why the first ionisation of elements vary going across a period and down a group

››› the link between successive ionisation energies and electronic structure

The process of removing electrons from an atom is called ionisation. The energy needed to remove each successive electron from an atom is called the first, second, third, etc., ionisation energy.

The process for the **first ionisation energy** (I.E.) of an element is summarised in the equation:

$$X(g) \longrightarrow X^+(g) + e^-$$

Electrons are held in their shells by their attraction to the positive nucleus, therefore the greater the attraction, the greater the ionisation energy. This attraction depends on three factors:

- The size of the positive nuclear charge – the greater the nuclear charge, the greater the attractive force on the outer electron and the greater the ionisation energy.

- The distance of the outer electron from the nucleus – the force of attraction between the nucleus and the outer electron decreases as the distance between them increases. The further an electron is from the nucleus, the lower the ionisation energy.

- The **shielding effect** by electrons in filled inner shells – all electrons repel each other since they are negatively charged. Electrons in the filled inner shells repel electrons in the outer shell and reduce the effect of the positive nuclear charge. The more filled inner shells or subshells there are, the smaller the attractive force on the outer electron and the lower the ionisation energy.

Evidence for shells and subshells can be seen from a plot of first ionisation energies against the elements (or atomic number). Such a plot is shown below for the first twenty elements.

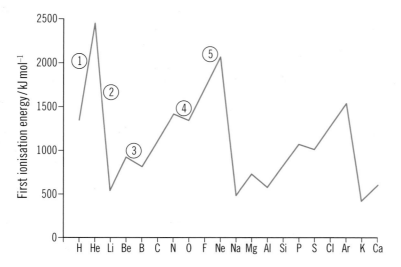

▲ Plot of I.E. against elements

The most significant features of the plot are:

- The 'peaks' are occupied by elements of Group 0.
- The 'troughs' are occupied by elements of Group 1.
- There is a general increase in ionisation energy across a period, although this increase is not uniform.

Looking at the plot in detail (remember the three main factors that affect ionisation energy):

1. He > H since helium has a greater nuclear charge in the same subshell so little extra shielding.

2. He > Li since lithium's outer electron is in a new shell which has increased shielding and is further from the nucleus.

3. Be > B since boron's outer electron is in a new subshell of slightly higher energy level and is partly shielded by the 2s electrons.

4. N > O since the electron–electron repulsion between the two paired electrons in one p orbital in oxygen makes one of the electrons easier to remove. Nitrogen does not contain paired electrons in its p orbital.

5. He > Ne since neon's outer electron has increased shielding from inner electrons and is further from the nucleus.

Key Terms

The **molar first ionisation energy** of an element is the energy required to remove one mole of electrons from one mole of its gaseous atoms.

Electron shielding or screening is the repulsion between electrons in different shells. Inner shell electrons repel outer shell electrons

Link Trends across periods page 57

▼ Study point

If the conditions for ionisation energy are 298 K and 1 atm then the process is known as the standard ionisation energy.

All ionisation energies are positive since it always requires energy to remove an electron.

Key Term

Successive ionisation energies are a measure of the energy needed to remove each electron in turn until all the electrons are removed from an atom.

14

Knowledge check

The first four ionisation energies (in $kJ\,mol^{-1}$) for an element are: 738, 1451, 7733 and 10541.

The element belongs to group .... in the periodic table because there is a .... between the .... and .... ionisation energies.

Stretch & Challenge

For any positive number n, log10 of n is the power to which the base (in this case 10) must be raised to make n. For example, for the number 100,

$Log_{10}\,100 = 2$ i.e. $100 = 10^2$

Exam tip

A large increase in successive ionisation energies shows that an electron has been removed from a new shell closer to the nucleus and gives the group to which the element belongs.

Li has a large energy jump between 1st and 2nd I.E. therefore it's in Group 1.

Al has a large energy jump between 3rd and 4th I.E. therefore it's in Group 3.

YOU SHOULD KNOW ›››

››› the relationship between energy and frequency and that of frequency and wavelength

››› the order of increasing energy between infrared, visible and ultraviolet light

Successive ionisation energies

Further evidence for shells and subshells comes from the **successive ionisation energies** needed to remove all the electrons from an atom.

An element has as many ionisation energies as it has electrons. Sodium has eleven electrons and so has eleven successive ionisation energies.

For example, the third ionisation energy is a measure of how easily a 2+ ion loses an electron to form a 3+ ion. An equation to represent the third ionisation energy of sodium is:

$$Na^{2+}(g) \longrightarrow Na^{3+}(g) + e^-$$

Successive ionisation energies always increase because:

- There is a greater effective nuclear charge as the same number of protons are holding fewer and fewer electrons.
- As each electron is removed there is less electron–electron repulsion and each shell will be drawn in slightly closer to the nucleus.
- As the distance of each electron from the nucleus decreases, the nuclear attraction increases.

As the ionisation energies are so large we must use logarithms to base 10 (log_{10}) to make the numbers fit on a reasonable scale.

The graph below shows the successive ionisation energies of sodium.

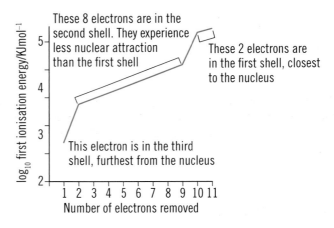

▲ Graph of sodium's I.E.

For sodium there is one electron on its own which is easiest to remove. Then there are eight more electrons which become successively more difficult to remove. Finally there are two electrons which are the most difficult to remove.

Notice the large increases in ionisation energy as the 2nd and 10th electrons are removed. If the electrons were all in the same shell, there would be no large rise or jump.

Emission and absorption spectra

Light and electromagnetic radiation

Light is a form of electromagnetic radiation. Electromagnetic radiation is energy travelling as waves. A wave is described by its frequency (f) and its wavelength (λ).

The frequency and wavelength of light are related by the equation:

$c = f\lambda$ (c is the speed of light)

The frequency of electromagnetic radiation and energy (*E*) are connected by the equation:

$E = hf$ (h is Planck's constant) — 6.6×10^{-39}

Therefore, $f \propto E$, and if frequency increases energy increases.

$f \propto 1/\lambda$ and if frequency increases wavelength decreases.

The whole range of frequencies of electromagnetic radiation is called the electromagnetic spectrum.

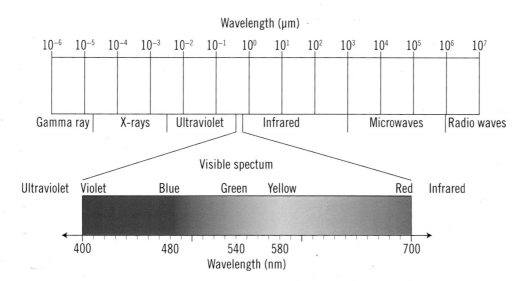

▲ The electromagnetic spectrum

In this unit we are only concerned with the infrared, visible and ultraviolet regions.

It is important to note that both energy and frequency increase going from the infrared through visible to the ultraviolet region. Thus, blue light is of a higher energy than red light. As frequency gets larger and therefore wavelength decreases, blue light must have a shorter wavelength than red light.

Absorption spectra

Light of all visible wavelengths is called white light. All atoms and molecules absorb light of certain wavelengths. Therefore, when white light is passed through the vapour of an element, certain wavelengths will be absorbed by the atoms and removed from the light. Looking through a spectrometer, black lines appear in the spectrum where light of some wavelengths has been absorbed. The wavelengths of these lines correspond to the energy taken in by the atoms to promote electrons from lower to higher energy levels.

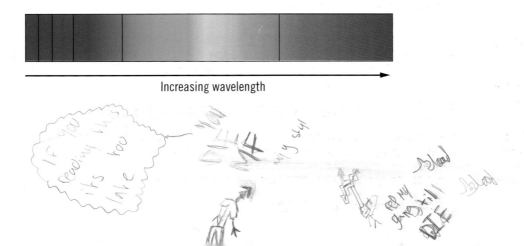

Increasing wavelength

Exam tip

Since $f \propto E$ and $f \propto 1/\lambda$, the lower the wavelength the higher the frequency and the greater the energy.

▼ **Study point**

Light is electromagnetic radiation in the range of wavelength corresponding to the visible region of the electromagnetic spectrum.

! **Extra Help**

Wavelength is the distance over which the wave's shape repeats.

Frequency (in Hz) is the number of times the wave is repeated in one second.

15 Knowledge check

Two lines in the emission spectrum of atomic hydrogen have the following frequencies 460 THz and 690 THz. State which one has the higher
(a) energy   (b) wavelength.

Emission spectra

When atoms are given energy by heating or by an electrical field, electrons are excited and the additional energy promotes them from a lower energy level to a higher one. When the source of energy is removed and the electrons leave the excited state, they fall from the higher energy level to a lower energy level and the energy lost is released as a photon (a quantum of light energy) with a specific frequency. The observed spectrum consists of a number of coloured lines on a black background.

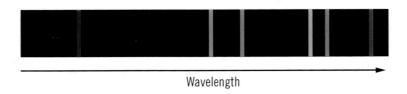

Wavelength

The fact that only certain colours appear in an atom's emission spectrum means that only photons having certain energies are emitted by the atom.

If the electron energy levels were not quantised but could have any value, a continuous spectrum rather than a line spectrum would result.

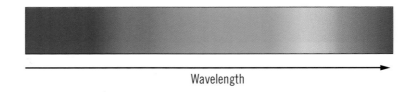

Wavelength

The hydrogen spectrum

An atom of hydrogen has only one electron so it gives the simplest emission spectrum. The atomic spectrum of hydrogen consists of separate series of lines mainly in the ultraviolet, visible and infrared regions of the electromagnetic spectrum. There are six series, each named after their discoverer. Only one series, the Balmer series, is in the visible region of the spectrum.

Diagram showing part of emission spectrum of atomic hydrogen.

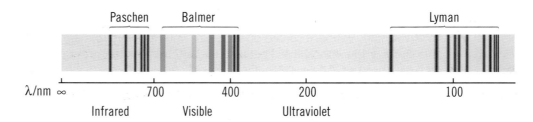

When an atom is excited by absorbing energy, an electron jumps up to a higher energy level. As the electron falls back down to a lower energy level it emits energy in the form of electromagnetic radiation. The emitted energy can be seen as a line in the spectrum because the energy of the emitted radiation is equal to the difference between the two energy levels, ΔE, in this electronic transition, i.e. it is a fixed quantity or quantum.

Since $\Delta E = hf$, electronic transitions between different energy levels result in emission of radiation of different frequencies and therefore produce different lines in the spectrum.

As the frequency increases, the lines get closer together because the energy difference between the shells decreases. Each line in the Lyman series (ultraviolet region) is due to electrons returning to the first shell or $n = 1$ energy level, while the Balmer series (visible region) is due to electrons returning to the second shell or $n = 2$ energy level.

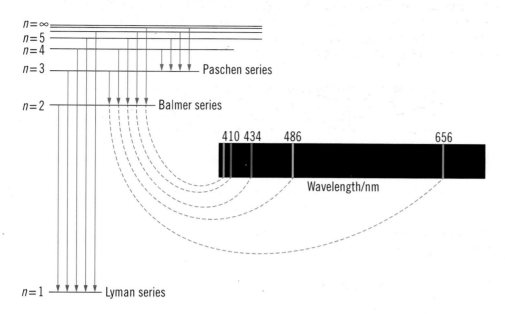

16

Knowledge check

Which letter represents the transition that causes a line of the lowest frequency in the emission spectrum of atomic hydrogen?

A   Second energy level to first energy level.

B   Third energy level to first energy level.

C   Third energy level to second energy level.

D   Fourth energy level to second energy level.

▼ **Study point**

The various series in the infrared region are caused by electrons returning to the energy levels $n = 3$ (Paschen), $n = 4$ (Brackett), $n = 5$ (Pfund) and $n = 6$ (Humphreys).

Ionisation of the hydrogen atom

The spectral lines become closer and closer together as the frequency of the radiation increases until they converge to a limit. The convergence limit corresponds to the point at which the energy of an electron is no longer quantised. At that point the nucleus has lost all influence over the electron; the atom has become ionised.

For the Lyman series, $n = 1$, the convergence limit represents the ionisation of the hydrogen atom.

Measuring the convergent frequency (difference from $n = 1$ to $n = \infty$) allows the ionisation to be calculated using $\Delta E = hf$.

The value of ΔE is multiplied by Avogadro's constant to give the first ionisation energy for a mole of atoms.

▼ **Study point**

The convergence limit is when the spectral lines become so close together they have a continuous band of radiation and separate lines cannot be distinguished.

Exam tip

The I.E. of a hydrogen atom can be shown on its electron energy level diagram by drawing an arrow upwards from the $n = 1$ to the $n = \infty$ level.

Worked example

The value of the wavelength at the start of the continuum in the hydrogen emission spectrum is 92 nm. Calculate the first ionisation energy of hydrogen.

(Assume that $c = 3.00 \times 10^8 \, m\,s^{-1}$, $h = 6.63 \times 10^{-34} \, Js$ and $L = 6.02 \times 10^{23} \, mol^{-1}$)

Ionisation Energy $= L\Delta E$ (L = Avogadro's constant)

But $\Delta E = hf$ and $f = c/\lambda$ (h = Planck's constant, c = speed of light)

Therefore Ionisation Energy $= Lhc/\lambda$

$$= 6.02 \times 10^{23} \times 6.63 \times 10^{-34} \times \frac{3.00 \times 10^8}{(92 \times 10^{-9})}$$

$$= 1\,301\,498 \, J\,mol^{-1}$$

$$= 1301 \, kJ\,mol^{-1}$$

17

Knowledge check

The value of the frequency at the start of the continuum in the sodium emission spectrum is 1.24×10^{15} Hz. Calculate the first ionisation of sodium.

Unit 1

1.3
Chemical calculations

Chemists are interested in how atoms react together. They want to know about the qualitative and quantitative aspects of chemical reactions. Very often chemists want to measure out exact quantities of substances that will react together – especially in industry where adding too much of a reagent will result in an unnecessary cost or may contaminate the product. What is really useful is to be able to work out these quantities on the basis of the number of atoms (or molecules) of substances that react together. This unit shows how this can be done by using the concept of the mole and balanced chemical equations.

Content

You should be able to demonstrate and apply knowledge and understanding of:

- The various relative mass terms (atomic, isotopic, formula, molecular).
- The principles of the mass spectrometer and its use in determining relative atomic mass and relative abundance of isotopes.
- Simple mass spectra, for example that of chlorine gas.
- How empirical and molecular formulae can be determined from given data.
- The relationship between the Avogadro constant, the mole and molar mass.
- The relationship between grams and moles.
- The concept of concentration and its expression in terms of grams or moles per unit volume (including solubility).
- Molar volume and correction due to changes in temperature and pressure.
- The ideal gas equation ($pV = nRT$).
- The concept of stoichiometry and its use in calculating reacting quantities, including in acid–base titrations.
- The concepts of atom economy and percentage yield.
- How to estimate the percentage error in a measurement and use this to express numeric answers to a sensible number of significant figures.

Relative mass terms

Masses of atoms

The masses of individual atoms are too small to be used in calculations in chemical reactions, so instead the mass of an atom is expressed relative to a chosen standard atomic mass. The carbon-12 isotope is taken as the standard of reference because relative atomic masses are determined by mass spectrometry and volatile carbon compounds are widely used in mass spectrometry.

Most elements exist naturally as two or more different isotopes. The mass of an element therefore depends on the relative abundance of all the isotopes present in the sample. In order to overcome this, chemists use an average mass of all the atoms and this is called the **relative atomic mass, A_r**

Relative atomic mass has no units since it is one mass compared with another mass.

If we refer to the mass of a particular isotope then the term **relative isotopic mass** is used.

Masses of compounds

Since the formula of a compound shows the ratio in which the atoms combine, the idea of relative atomic mass can be extended to compounds and the term **relative formula mass, M_r**, is used.

E.g. the relative formula mass of copper(II) sulfate, $CuSO_4$ is:

$$(1 \times 63.5) + (1 \times 32) + (4 \times 16) = 159.5$$

The mass spectrometer

When a mass spectrometer is used to find the relative atomic mass of an element it measures two things:

- The mass of each different isotope of the element.
- The relative abundance of each isotope of the element.

The diagram below shows how a mass spectrometer works.

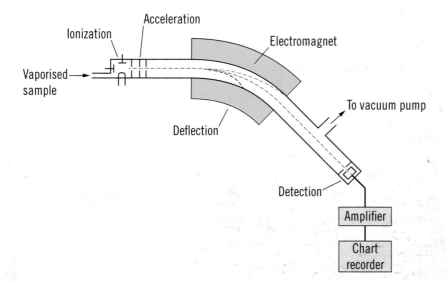

During the detecting stage, when an ion hits the metal box, its charge is neutralised by an electron jumping from the metal on to the ion. That leaves a space amongst the electrons in the metal, and the electrons in the wire shuffle along to fill it.

A flow of electrons in the wire is detected as an electric current which can be amplified and recorded. The more ions arriving, the greater the current.

There are four main steps:

Ionisation The vaporised sample passes into the ionisation chamber.

The particles in the sample (atoms or molecules) are therefore bombarded with a stream of electrons, and some of the collisions are sufficiently energetic to knock one or more electrons out of the sample particles to make positive ions.

Most of the positive ions formed will carry a charge of +1 because it is much more difficult to remove further electrons from an already positive ion.

Acceleration An electric field accelerates the positive ions to high speed.

Deflection Different ions are deflected by the magnetic field by different amounts. The amount of deflection depends on:

- The mass of the ion. Lighter ions are deflected more than heavier ones.
- The charge on the ion. Ions with two (or more) positive charges are deflected more than ones with only one positive charge.

These two factors are combined into the mass/charge ratio.

(Unless otherwise stated, mass spectra given will only involve 1+ ions, so the mass/charge ratio will be the same as the mass of the ion.)

Detection The beam of ions passing through the machine is detected electrically. Only ions with the correct mass/charge ratio make it right through the machine to the ion detector. (The other ions collide with the walls where they will pick up electrons and be neutralised. Eventually, they get removed from the mass spectrometer by the vacuum pump.) The signal is then amplified and recorded.

It's important that the ions produced in the ionisation chamber have a free run through the machine without hitting air molecules so a vacuum is needed inside the apparatus.

Calculating relative atomic mass

As stated, data from mass spectrometry can be used to calculate relative atomic mass, e.g. below is the mass spectrum of lead

19

Knowledge check

A mass spectrum of a sample of hydrogen showed that it contained ^1H 99.20% and ^2H 0.8000%.

Calculate the relative atomic mass of the hydrogen sample, giving your answer to **four** significant figures.

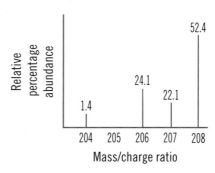

The relative atomic mass is a weighted average of the masses of all the atoms in the isotopic mixture, therefore

$$\text{Relative atomic mass} = \frac{(1.40 \times 204) + (24.1 \times 206) + (22.1 \times 207) + (52.4 \times 208)}{100}$$

$$= 207.2$$

Other uses of mass spectrometry include:

- Identifying unknown compounds, e.g. testing athletes for prohibited drugs.
- Identifying trace compounds in forensic science.
- Analysing molecules in space.

Interpretation of mass spectra

When a vaporised compound passes through a mass spectrometer, an electron is knocked off a molecule to form a positive ion. This ion is called the molecular ion and its mass gives the relative formula mass of the compound.

The molecular ions are energetically unstable, and some of them will break up into smaller pieces or fragments.

All sorts of fragmentations of the original molecular ion are possible – and that means that you will get a wide range of lines in the mass spectrum.

The mass spectrum of chlorine

Chlorine is made up of two isotopes ^{35}Cl and ^{37}Cl. Chlorine gas consists of molecules not individual atoms, but the mass spectrum of chlorine is:

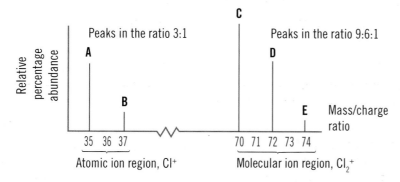

When chlorine is passed into the ionisation chamber, an electron is knocked off the molecule to give a molecular ion, Cl_2^+. These ions won't be particularly stable, and some will fall apart to give a chlorine atom and a Cl^+ ion. (This is known as fragmentation.)

So peak A is caused by $^{35}Cl^+$ and peak B by $^{37}Cl^+$.

As the ^{35}Cl isotope is three times more common than the ^{37}Cl isotope, the heights of the peaks are in the ratio of 3:1.

In the molecular ion region think about the possible combinations of ^{35}Cl and ^{37}Cl atoms in a Cl_2^+ ion. Both atoms could be ^{35}Cl, both atoms could be ^{37}Cl, or you could have one of each sort.

So peak C (*m/z* 70) is due to $(^{35}Cl—^{35}Cl)^+$

Peak D (*m/z* 72) is due to $(^{35}Cl—^{37}Cl)^+$ or $(^{37}Cl—^{35}Cl)^+$

Peak E (*m/z* 74) is due to $(^{37}Cl—^{37}Cl)^+$

Since the probability of an atom being ^{35}Cl is $\frac{3}{4}$ and that of being ^{37}Cl is $\frac{1}{4}$, then

molecule	$^{35}Cl—^{35}Cl$	$^{35}Cl—^{37}Cl$ or $^{37}Cl—^{35}Cl$	$^{37}Cl—^{37}Cl$
probability	$\frac{3}{4} \times \frac{3}{4}$	$\frac{3}{4} \times \frac{1}{4}$ or $\frac{1}{4} \times \frac{3}{4}$	$\frac{1}{4} \times \frac{1}{4}$
	$\frac{9}{16}$	$\frac{6}{16}$	$\frac{1}{16}$

and ratio of peaks C : D : E is 9 : 6 : 1

The molecular ion region of mass spectra can give information about the isotopes in the molecule.

Bromine has two isotopes. The molecular ion region of its mass spectrum is shown on the right.

YOU SHOULD KNOW ›››

››› how to interpret simple mass spectra

‹Link› Mass spectrometry page 161

▼ **Study point**

For the chlorine spectrum you can't make any predictions about the relative heights of the lines at *m/z* 35/37 compared with those at 70/72/74. That depends on what proportion of the molecular ions break up into fragments.

20

Knowledge check

In the mass spectrum of hydrogen, explain why peaks due to hydrogen atoms are present although hydrogen gas contains only H_2 molecules.

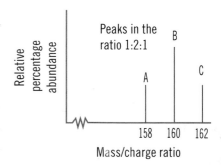

Since it only has two isotopes:

peak A (m/z 158) must be due to $(^{79}Br—^{79}Br)^+$ and

peak C (m/z 162) must be due to $(^{81}Br—^{81}Br)^+$.

Therefore peak B (m/z 160) is due to $(^{79}Br—^{81}Br)^+$ or $(^{81}Br—^{79}Br)^+$.

Since the ratio of peaks A : B : C is almost 1 : 2 : 1 it follows that natural bromine consists of a nearly 50:50 mixture of ^{79}Br and ^{81}Br.

molecule	$^{79}Br—^{79}Br$	$^{79}Br—^{81}Br$ or $^{81}Br—^{79}Br$	$^{81}Br—^{81}Br$
probability	$\frac{1}{2} \times \frac{1}{2}$	$\frac{1}{2} \times \frac{1}{2}$ or $\frac{1}{2} \times \frac{1}{2}$	$\frac{1}{2} \times \frac{1}{2}$
	$\frac{1}{4}$	$\frac{2}{4}$	$\frac{1}{4}$

Amount of substance

In chemical reactions, for all the reactants to change into products, the correct quantity of each reactant must be used. Since the atoms that make up the reactants rearrange to form the products, it would be really useful to be able to work out these quantities on the basis of the number of atoms that react.

If you pay coins into a bank, the clerk won't count them individually – he knows the mass of a certain number of coins, e.g. £10 worth of 50p, so he will weigh a bag of coins to confirm that the correct number of coins is present.

Similarly, atoms are too small to be counted individually, so they are counted by weighing a collection of them where the mass of a particular fixed number of atoms is known.

Since carbon-12 is the standard chosen for relative atomic mass, the number of atoms in exactly 12 g of carbon-12 is chosen as the standard and is known as the **mole**.

The number of atoms in a mole is very large. Using a mass spectrometer, the mass of a single carbon-12 atom has been found to be 1.993×10^{-23} g therefore

$$\text{the number of atoms per mole} = \frac{\text{mass per mole of } ^{12}C}{\text{mass of one atom of } ^{12}C}$$

$$= \frac{12\,g\,mol^{-1}}{1.993 \times 10^{-23}\,g}$$

$$= 6.02 \times 10^{23}\,mol^{-1}$$

This is called the **Avogadro constant**, **L**, after the nineteenth-century Italian chemist Amadeo Avogadro.

The mass of one mole of a substance is called the **molar mass**, **M**. It has the same numerical value as A_r or M_r but has the unit $g\,mol^{-1}$.

The amount of substance in moles (n), the mass (m) and molar mass (M) are linked by the equation:

$$\text{amount in moles } (n) = \frac{\text{mass } (m)}{\text{molar mass } (M)}$$

The expression for amount in moles may be rearranged in two ways to find the mass of a sample or molar mass of a substance.

Worked examples

What is the amount of sodium present in 0.23 g of sodium?

A_r of sodium = 23.0

$$\text{Molar mass of sodium} = 23.0\,\text{g mol}^{-1}$$

$$\text{Amount of sodium} = \frac{m}{M} = \frac{0.23\,\text{g}}{23.0\,\text{g mol}^{-1}} = 0.01\,\text{mol}$$

If you need 0.05 mol of sodium hydroxide, what mass of the substance do you have to weigh out?

$$\text{Molar mass of NaOH} = 40\,\text{g mol}^{-1}$$

$$\text{Mass of sample} = n \times M$$

$$= 0.05 \times 40$$

$$= 2\,\text{g}$$

Calculating reacting masses

In a chemical reaction, reactants change into products. If we are given the mass of reactants, we can find out the mass of products formed as long as we have a balanced chemical equation for the reaction.

An equation tells us not only what substances react together but also what amounts of substances in moles react together. The ratio between amounts in moles of reactants and products are called the **stoichiometric** ratios (mole ratios).

E.g. the equation for the burning of magnesium in oxygen to produce magnesium oxide

$$2Mg + O_2 \longrightarrow 2MgO$$

tells us that 2 moles of magnesium react with 1 mole of oxygen to produce 2 moles of magnesium oxide.

What mass of magnesium oxide forms if we burn 1.215 g of magnesium in an excess of oxygen?

To calculate this, the following route is used:

Step 1   Change the mass of Mg into amount of moles (divide by molar mass).

Step 2   Use the balanced equation to state the mole ratio of Mg:MgO, hence deduce the moles of MgO.

Step 3   Change the amount in moles of MgO to mass (multiply by molar mass).

Step 1   Amount in moles of Mg $= \dfrac{1.215}{24.3} = 0.050\,\text{mol}$

Step 2   The mole ratio from the equation is 2Mg:2MgO i.e. 1:1

therefore 0.050 mol Mg gives 0.050 mol MgO

Step 3   Molar mass of MgO $= 24.3 + 16 = 40.3$

Mass MgO $= 0.05 \times 40.3$

$= 2.015\,\text{g}$

21

Knowledge check

Calculate the mass of sodium that contains

(a) 6×10^{23} atoms

(b) 4×10^{25} atoms

(Assume that the Avogadro constant is $6 \times 10^{23}\,\text{mol}^{-1}$)

22

Knowledge check

(a) Calculate the mass of 0.020 mol of sodium carbonate.

(b) Calculate the amount (in moles) of 1.36 g calcium carbonate.

Key Term

Stoichiometry is the molar relationship between the amounts of reactants and products in a chemical reaction.

When iron ore (iron(III) oxide) is reduced by carbon monoxide in a blast furnace, iron is produced. Calculate how much iron is produced from 1 kg of iron ore.

▼ Study point

Excess oxygen implies that there is more than enough oxygen present to react with all the magnesium burnt in it. When the reaction is complete, all the magnesium has reacted and some oxygen remains.

Key Terms

Empirical formula is the simplest formula showing the simplest whole number ratio of the amount of elements present.

Molecular formula shows the actual number of atoms of each element present in the molecule. It is a simple multiple of the empirical formula.

▼ Study point

When you divide the percentages by the relevant atomic masses, do not truncate the answers, e.g. 1.25 to 1. The figures provided in the question should give reasonably simple ratios for the empirical formula.

23

Knowledge check

Find the empirical formula of the compound formed when 1.172 g iron forms 3.409 g of a chloride.

Empirical and molecular formulae

We can use calculations involving masses of the elements that combine together to find the formulae of compounds.

Empirical formula is the simplest formula showing the simplest whole number ratio of the amount of elements present.

Molecular formula shows the actual number of atoms of each element present in the molecule. It is a simple multiple of the empirical formula. Usually the relative formula mass is needed to determine the molecular formula.

Empirical formulae can be calculated from known masses or percentage composition data. There are three steps in the calculation:

Step 1   Find the amount in moles of each element present (divide by the molar mass).

Step 2   Find the ratio of the number of atoms present (divide by the smallest value in step 1).

Step 3   Convert these numbers into whole numbers (atoms combine together in whole number ratios).

Worked example

A compound of carbon, hydrogen and oxygen has a relative molecular mass of 60. The percentage composition by mass is C 40.0%; H 6.70%; O 53.3%.

What is (a) the empirical formula and (b) the molecular formula?

(a)

	C	:	H	:	O
Molar ratio of atoms	$\dfrac{40}{12}$		$\dfrac{6.7}{1.01}$		$\dfrac{53.3}{16}$
	3.33		6.63		3.33
Divide by smallest number	1		2		1

Empirical formula is CH_2O

(b) Mass of empirical formula $= 12 + 2.02 + 16 = 30.02$

Number of CH_2O units in a molecule $= \dfrac{60}{30.02} = 2$

Molecular formula is $C_2H_4O_2$

Some salts have water molecules incorporated into their structure. These are known as hydrated salts and the water is known as water of crystallisation. If we know the mass of the anhydrous salt and the mass of the water in the hydrated salt we can calculate the number of moles of water in the hydrated salt.

Worked example

Sodium carbonate can form a hydrate, $Na_2CO_3.xH_2O$. When 4.64 g of this hydrate was heated, 2.12 g of the anhydrous salt, Na_2CO_3, remained.

What is the value of x?

Mass of water in the hydrate $= 4.64 - 2.12 = 2.52$ g

Moles $Na_2CO_3 = \dfrac{2.12}{106} = 0.020$

Moles $H_2O = \dfrac{2.52}{18.02} = 0.140$

Mole ratio of H_2O : Na_2CO_3

0.140 : 0.020

Divide by smaller number   7   :   1

Value of $x = 7$ and formula is $Na_2CO_3.7H_2O$

Volumes of gases

For reactions involving gases, it is more usual to consider the volumes of reactants and products rather than their masses. In 1811, Avogadro discovered that equal volumes of all gases contain the same number of molecules. (Volumes must be measured under the same conditions of temperature and pressure.) This provides a way of calculating the amount of gas present in a given volume. At standard temperature and pressure (stp), 0°C and 1 atm., one mole of any gas occupies $22.4 \, dm^3$. This is known as the gas **molar volume**, v_m.

The steps in calculations involving molar volume are similar to the ones for reacting masses.

E.g. What volume of hydrogen is produced, at stp, where 3.00 g of zinc reacts with excess hydrochloric acid?

(1 mole of hydrogen occupies $22.4 \, dm^3$ at stp)

$$Zn + 2HCl \longrightarrow ZnCl_2 + H_2$$

Step 1   Change the mass of zinc into moles

Amount of moles of zinc $= \dfrac{3.00}{65.4} = 0.0459$

Step 2   The mole ratio from the equation is $1Zn : 1H_2$

therefore $0.0459 \, mol$ Zn gives $0.0459 \, mol \, H_2$

Step 3   Change the moles into volume of gas

Volume of hydrogen $= 0.0459 \times 22.4 = 1.03 \, dm^3$

The gas laws

There are three important ideas describing the behaviour of gases: Boyle's law, Charles' law and Avogadro's principle.

Boyle's law

In 1662, Robert Boyle published his work on the compressibility of gases. He stated that:

At a constant temperature, the volume of a fixed mass of gas is inversely proportional to its pressure.

It can be written as $V \propto 1/P$ or $PV = $ constant.

Graphs illustrating Boyle's law are shown below:

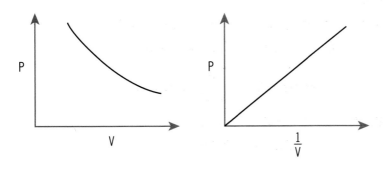

Charles' law

In 1787 Jacques Charles stated his law on the effect of temperature on the volume of a gas. He stated that:

> The volume of a fixed mass of a given gas, at constant pressure, is directly proportional to its temperature in kelvins.

It can be written as $V \propto T$ or $\dfrac{V}{T} = $ constant

A graph illustrating Charles' law is shown below:

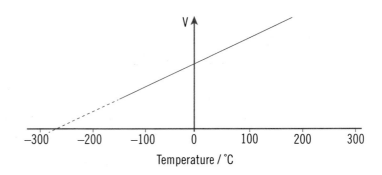

Avogadro's principle

In 1811 Amadeo Avogadro proposed the following hypothesis about gases:

> Equal volumes of different gases, measured at the same temperature and pressure, contain the same number of molecules.

Another way of putting this is that the volume of a gas depends on the amount of moles or $V \propto n$

The ideal gas equation

Combining Boyle's law and Charles' law gives

$$PV \propto T \quad \text{or} \quad \frac{PV}{T} = \text{constant}$$

It follows from Avogadro's hypothesis that, since volume is proportional to the amount of moles, if one mole of gas is considered, the constant will be the same for all gases.

This constant is called the gas constant and is given the symbol R.

Therefore for n moles of gas $\quad \dfrac{PV}{T} = nR$

or $\qquad\qquad\qquad\qquad PV = nRT$

This is known as the ideal gas equation and R has the value $8.314\,J\,K^{-1}\,mol^{-1}$.

In calculations involving the ideal gas equation SI units must be used, i.e.

Pressure must be in Pa (pascals)

Volume must be in m^3

Temperature must be in K (kelvins)

Worked example

A sample of sulfur dioxide occupied a volume of $12.0\,dm^3$ at $10.0\,°C$ and a pressure of $105\,kPa$.

Calculate the amount, in moles, of sulfur dioxide in this sample.

(The gas constant $R = 8.31\,J\,K^{-1}\,mol^{-1}$)

$$PV = nRT$$

Therefore

$$n = \frac{PV}{RT}$$

SI units must be used:

$P = 105000\,Pa$ (105×1000)

$V = 0.012\,m^3$ $(12/1000)$

$T = 283\,K$ $(10 + 273)$

$$n = \frac{105000 \times 0.012}{8.31 \times 283}$$

$$n = 0.536\,mol$$

We can also use the ideal gas equation to calculate the volume a gas would occupy at temperatures and pressures other than those at which it was actually measured.

If a gas has a volume V_1 measured at pressure P_1 and temperature T_1, then:

$$\frac{P_1 V_1}{T_1} = nR$$

If the same gas's volume V_2 is now measured at a new pressure P_2 and temperature T_2, then:

$$\frac{P_2 V_2}{T_2} = nR$$

Therefore

$$\frac{P_1 V_1}{T_1} = \frac{P_2 V_2}{T_2}$$

Worked example

A sample of gas collected at $30.0\,°C$ and $1.01 \times 10^5\,N\,m^{-2}$ had a volume of $40.0\,cm^3$. What is the volume of the gas at standard temperature and pressure (stp)?

(At stp, pressure = $1.01 \times 10^5\,N\,m^{-2}$ and temperature = $273\,K$)

$$\frac{P_1 V_1}{T_1} = \frac{P_2 V_2}{T_2}$$

$P_1 = 1.01 \times 10^5\,N\,m^{-2}$ $\qquad P_2 = 1.01 \times 10^5\,N\,m^{-2}$

$V_1 = 40.0\,cm^3$ $\qquad V_2 = ?$

$T_1 = 30\,°C = 303\,K$ $\qquad T_2 = 273\,K$ (Temperature must be in kelvins)

$$\frac{1.01 \times 10^5 \times 40}{303} = \frac{1.01 \times 10^5 \times V_2}{273}$$

$$V_2 = \frac{1.01 \times 10^5 \times 40 \times 273}{1.01 \times 10^5 \times 303}$$

$$V_2 = 36.0\,cm^3$$

Concentrations of solutions

The concentration of a solution measures how much of a dissolved substance is present per unit volume of a solution. A solution with a large quantity of solute in a small quantity of solvent is described as concentrated. A solution with a small quantity of solute in a large quantity of solvent is described as dilute.

Study point

In calculations involving $\frac{P_1 V_1}{T_1} = \frac{P_2 V_2}{T_2}$ remember to change temperature to kelvin.

25 Knowledge check

State the units that must be used for pressure, volume and temperature in the ideal gas equation.

26 Knowledge check

Calculate the volume, in dm^3, occupied by $0.10\,mol$ of carbon dioxide at $1.01 \times 10^5\,N\,m^{-2}$ and $110\,°C$.

(Gas constant $R = 8.31\,J\,K^{-1}\,mol^{-1}$)

27 Knowledge check

Calculate the volume of a gas at stp if it occupies $200\,cm^3$ at $50\,°C$ and $2\,atm$.

(Conditions at stp are $273\,K$ and $1\,atm$)

YOU SHOULD KNOW ›››

››› how to calculate concentration in terms of grams or moles per unit volume

▼ Study point

Solute is the dissolved substance. The liquid in which the solute dissolves is the solvent (usually water)

Extra Help

Remember 1 litre = $1\,dm^3$ = $1000\,cm^3$.

$1\,cm^3$ = 1 ml.

To change cm^3 into dm^3 you must divide by 1000

Stretch & Challenge

The concentration of magnesium ions in a sample of Welsh mineral water is 15mg/litre. Calculate the concentration in $mol\,dm^{-3}$.

▼ Study point

Learn the equations that connect amount of substance and concentration for a solution:

$n = cv$ or $c = \dfrac{n}{V}$ or $v = \dfrac{n}{c}$

Remember if v is given in cm^3 divide by a 1000 to change it into dm^3.

28

Knowledge check

Calculate the concentration of 0.037 g of calcium hydroxide in $200\,cm^3$ of solution.

YOU SHOULD KNOW ›››

››› how to calculate reacting quantities in acid–base titrations

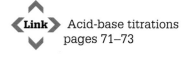

Link Acid-base titrations pages 71–73

The concentration of a solution can be stated in many ways, e.g. the concentration of ions in a bottle of mineral water is stated in mg/litre, but the most convenient way is to state the amount in moles of a solid present in $1\,dm^3$ of solution.

i.e. concentration = $\dfrac{\text{amount of moles of solute}}{\text{volume of solution}}$ or $c = \dfrac{n}{V}$

and the unit is moles per cubic decimetre or $mol\,dm^{-3}$.

This expression can be rearranged in two ways to find the amount in moles or the volume,

e.g. 2.1 g of sodium hydrogencarbonate, $NaHCO_3$, is dissolved in $250\,cm^3$ of water. What is the concentration of the solution in $mol\,dm^{-3}$?

$$\text{Molar mass of } NaHCO_3 = 84$$

$$\text{Amount of moles } NaHCO_3 = \frac{2.1}{84} = 0.025$$

$$\text{Volume of water is } 250\,cm^3 = \frac{250}{1000} = 0.25\,dm^3$$

$$\text{Concentration} = \frac{0.025}{0.25} = 0.100\,mol\,dm^{-3}$$

Another way of expressing concentration is by using solubility. Solubility is usually measured in g/100g water and since the density of water is $1\,g\,cm^{-3}$, this is the same as $g/100\,cm^3$. Therefore to change solubility into concentration in $mol\,dm^{-3}$ the following steps are used.

Step 1 Change mass into moles (divide by molar mass)

Step 2 Change $100\,cm^3$ into dm^3 (divide by 1000)

Step 3 Divide the moles by the volume.

E.g. The solubility of potassium nitrate, KNO_3, is 31.6 g/100g water. What is the concentration in $mol\,dm^{-3}$?

Step 1 $\text{Molar mass } KNO_3 = 101.1$

$\text{Amount in moles} = \dfrac{31.6}{101.1} = 0.313$

Step 2 $100\,cm^3 \text{ water} = 0.100\,dm^3$

Step 3 $\text{Concentration} = \dfrac{0.313}{0.100} = 3.13\,mol\,dm^{-3}$

Acid–base titration calculation

Volumetric analysis is a means of finding the concentration of a solution. An acid–base titration is a type of volumetric analysis. In an acid–base titration, the reacting volumes of an acidic solution and a basic solution are accurately determined. If the concentration of one of these solutions is known, by using the stoichiometric ratios of the solutions, the concentration of the unknown solution can be determined.

Again, there are three main steps to follow:

Step 1 Find the amount of moles of the solution for which you know the concentration (e.g. acid).

Step 2 Use a balanced chemical equation to give the stoichiometric (mole) ratio between the acid and base.

Step 3 Calculate the concentration of the second solution (e.g. base) from the known volume and amount in moles.

Worked example

A 25.0 cm³ sample of aqueous sodium hydroxide was exactly neutralised by 21.0 cm³ of 0.150 mol dm⁻³ sulfuric acid.

Calculate the concentration of the sodium hydroxide solution in (a) mol dm⁻³ (b) g dm⁻³.

$$H_2SO_4 + 2NaOH \longrightarrow Na_2SO_4 + 2H_2O$$

Step 1 Amount in moles of H_2SO_4 = $0.150 \times 0.021 = 3.15 \times 10^{-3}$ mol

(Divide 21.0 cm³ by 1000 to change it into dm³)

Step 2 From the equation 1 mol H_2SO_4 requires 2 mol NaOH

3.15×10^{-3} mol H_2SO_4 require 6.30×10^{-3} mol NaOH

Step 3 (a) Concentration of NaOH = $\dfrac{6.30 \times 10^{-3}}{0.025}$ = 0.252 mol dm⁻³

(Divide 25.0 cm³ by 1000 to change it into dm³)

(b) M_r NaOH = 40.0

Concentration NaOH = $40.0 \times 0.252 = 10.1$ g dm⁻³

In a back-titration, the amount of excess reactant (e.g. acid) unused at the end of a reaction is found and so the amount used can be calculated. Using the stoichiometric ratios of the acid and base allows us to calculate the exact amount of the second reactant (e.g. base) that has reacted.

Worked example

A sample containing ammonium sulfate was warmed with 100 cm³ of 1.00 mol dm⁻³ sodium hydroxide solution. After all the ammonia had been evolved, the excess of sodium hydroxide solution was neutralised by 50.0 cm³ of 0.850 mol dm⁻³ hydrochloric acid.

What mass of ammonium sulfate did the sample contain?

The two reactions taking place are

(i) The reaction between the ammonium sulfate and sodium hydroxide

$$(NH_4)_2SO_4(s) + 2NaOH(aq) \longrightarrow 2NH_3(g) + Na_2SO_4(aq) + 2H_2O(l)$$

(ii) The neutralisation of the sodium hydroxide

$$NaOH(aq) + HCl(aq) \longrightarrow NaCl(aq) + H_2O(l)$$

Step 1 Calculate the amount in moles of HCl used in the neutralisation, this will give the amount in moles of NaOH unused in reaction (i)

Amount in moles of HCl = $0.850 \times \dfrac{50.0}{1000} = 0.0425$ mol

From equation (ii) mole ratio of HCl : NaOH is 1 : 1

Therefore amount of moles NaOH unused is 0.0425

Step 2 Calculate the amount in moles of NaOH that reacted with the $(NH_4)_2SO_4$

Initial amount of NaOH = $1.00 \times \dfrac{100}{1000} = 0.100$ mol

Amount in moles of NaOH used in reaction (i)
= $0.100 - 0.0425 = 0.0575$ mol

Step 3 Calculate the amount in moles of $(NH_4)_2SO_4$ used

From equation (i) mole ratio of NaOH : $(NH_4)_2SO_4$ is 2 : 1

Therefore 0.0575 moles NaOH react with 0.0288 moles $(NH_4)_2SO_4$

 Extra Help

In acid–base titrations there are five things you need to know:

The volume of the acid solution.

The concentration of the acid solution.

The volume of the base solution.

The concentration of the base solution.

The equation for the reaction.

If you know four of these, you can calculate the fifth.

29

Knowledge check

25.0 cm³ of hydrochloric acid are neutralised by 18.5 cm³ of a 0.200 mol dm⁻³ solution of sodium hydroxide. Calculate the concentration of the acid.

Step 4 Calculate the mass of $(NH_4)_2SO_4$ in the sample

Molar mass of $(NH_4)_2SO_4$ is $132\,g\,mol^{-1}$

Mass of $(NH_4)_2SO_4 = 0.0288 \times 132 = 3.80\,g$

YOU SHOULD KNOW ›››

››› how to calculate atom economies and percentage yields

Key Terms

Atom economy is obtained from the chemical equation for the reaction.

Atom economy =
$$\frac{\text{mass of required product}}{\text{total mass of reactants}} \times 100\%$$

% yield is calculated from the mass of product actually obtained by experiment.

% yield =
$$\frac{\text{mass (or moles) of product obtained}}{\text{maximum theoretic mass (or moles)}}$$
$\times 100\%$

30

Knowledge check

Calculate the atom economy for the production of iron in the reduction of iron(III) oxide by carbon monoxide.

$Fe_2O_3(s) + 3CO(g)$
$\longrightarrow 2Fe(s) + 3CO_2(g)$

Atom economy and percentage yield

When a reaction occurs, the compounds formed, other than the product needed are a waste. An indication of the efficiency of a reaction can be given as its **atom economy** or its **percentage yield**.

The higher the atom economy the more efficient the process.

Worked examples

Titanium used to be manufactured from titanium(IV) oxide by converting titanium(IV) oxide to titanium(IV) chloride then reducing it by magnesium. The process can be represented by:

$$TiO_2 + 2Cl_2 + 2Mg + 2C \longrightarrow Ti + 2MgCl_2 + 2CO$$

Calculate the percentage yield for this reaction.

$$\text{Atom economy} = \frac{\text{mass of required product}}{\text{total mass of reactants}} \times 100$$

$$= \frac{47.9}{79.9 + 142 + 48.6 + 24} \times 100$$

$$= 16.3\%$$

$32.1\,g$ of ethanoic acid are obtained from the oxidation of $27.6\,g$ of ethanol. The reaction can be represented by the equation:

$$C_2H_5OH + 2[O] \longrightarrow CH_3CO_2H + H_2O$$

Calculate the percentage yield for this reaction.

$$\text{% yield} = \frac{\text{mass of product obtained}}{\text{maximum theoretical yield}} \times 100$$

To calculate the maximum mass the three steps on page 35 must be followed:

Step 1 Calculate the amount, in mol, of $27.6\,g$ C_2H_5OH

$$n = \frac{m}{M} = \frac{27.6}{46.0} = 0.600$$

Step 2 Use the equation to calculate the amount in moles of CH_3CO_2H formed

1 mol C_2H_5OH gives 1 mol CH_3CO_2H

0.600 mol C_2H_5OH give 0.600 mol CH_3CO_2H

Step 3 Calculate the mass of CH_3CO_2H

$$m = nM = 0.600 \times 60.0 = 36.0\,g$$

$$\text{% yield} = \frac{32.1}{36.0} \times 100 = 89.2\%$$

Percentage error

Chemistry and all science depend on quantitative measurements in experiments. The results are expressed in three parts – a number, the units in which the number is given and an estimate of the error in the number. Care must be taken over the units, especially not to confuse J and kJ and g and kg but this section is about estimating the error.

Basically there are two ways of estimating the error.

If several measurements of a quantity can be made the values are averaged to obtain a mean value and a standard deviation is calculated.

If one or two measurements are taken (as is usually the case in AS chemistry), the error is estimated from the uncertainties in the equipment used, such as burette, pipette, balance and thermometer.

Remember that although these estimates are really 'guesstimates' they are important and essential.

Typically the error is taken as one-half of the smallest division on the apparatus, such as $0.05\,cm^3$ on a burette, $0.1°$ on a $0.2°$ thermometer and $0.5\,mg$ on a three-place balance. Remember that with most equipment, since the difference between the initial and final reading is required, two readings are taken so the errors will be $0.1\,cm^3$ for a burette, $0.2°$ for a $0.2°$ thermometer and $1.0\,mg$ $(0.001\,g)$ for a three-place balance.

These errors are now expressed as percentage errors by dividing by the quantity being measured, giving, for example.

burette – volume used $24.65\,cm^3$ (error $0.1\,cm^3$)

$$\% \text{ error } = \frac{0.10}{24.65} \times 100 = 0.406\%$$

thermometer – ΔT $7.0°$ (error $0.2°$)

$$\% \text{ error } = \frac{0.20}{7.0} \times 100 = 2.86\%$$

balance – $3.610\,g$ (error $0.002\,g$)

$$\% \text{ error } = \frac{0.002}{3.610} \times 100 = 0.055\%$$

Worked example

In an enthalpy of neutralisation reaction, the temperature of $100\,g$ of solution before the reaction was $18.6\,°C$ and after the reaction was $26.2\,°C$.

Calculate the percentage error caused by the thermometer and the error in the evaluation of the heat produced.

(The thermometer was accurate to $\pm 0.1\,°C$ and the specific heat capacity of the solution was $4.2\,J\,g^{-1}\,°C^{-1}$)

Temperature rise $= 26.2 - 18.6 = 7.6\,°C$

$$\% \text{ error } = \frac{0.20}{7.6} = 2.63\%$$

Heat produced $= 100 \times 4.2 \times 7.6 = 3192\,J$

Error in value of heat produced $= \pm 2.63\%$ of $3192\,J = \pm 83.9\,J$

Dealing with more than one percentage error

In a volumetric titration we may have percentage errors from the burette, pipette and balance contributing to the total error. Usually at AS it is simpler and satisfactory to identify

Stretch & Challenge

If several values have been obtained for a quantity, such as a titration value, these may be averaged to give a mean value, the spread or precision of the values found along with the percentage error and any dubious results rejected. This is a direct result from the experiment and does not need any estimate of burette accuracy, etc.

The following titration values were obtained in cm^3

22.6 23.1 22.9 23.9 22.5 22.4

The mean (average) of these is $\frac{137.4}{6} = 22.9$.

To obtain what is called the standard deviation we find the deviation of each value from the mean, such as -0.3 for 22.6, square each of them (giving $+0.09$ for our example), add up the squares, divide by the number of values (6) and take the square root.

This gives a standard deviation of 0.51 and we see that all of our values except 23.9 fall within the standard deviation; 23.9 is a long way out and we are entitled to reject it.

We can use standard deviations to compare the efficiency of competing processes or drugs. For example, drug A is found to have an effectiveness of 75% and drug B 77%.

Is B better than A or is it just chance? Standard deviations can give us the answer. If these are both 0.5% there is a 95% chance that the difference is real but if they are 1.5% there is a good chance that it is not.

The standard deviation is equivalent to the percentage error calculated using the method of estimating the uncertainties above. More than two-thirds of the measurements fall within one standard deviation of the mean.

▼ Study point

If an error is about three times larger than any other errors there is no need to consider the others at AS level.

31

Knowledge check

State the number of significant figures for the following numbers;

(a) 17.68   (b) 3.076   (c) 0.004
(d) 6.05×10^{18} (e) 3000.0

---

the largest error and use that on its own. For example, it is very rare for the percentage error in the balance to be significant.

The result of a calculation using several pieces of data cannot have more significant figures than the least number given in any of the terms in the calculation.

For example, solutions A and B react in a $1:1$ mole ratio and $10.00\,cm^3$ of solution B required $25.16\,cm^3$ of solution A the concentration of which was $0.10\,mol\,dm^{-3}$.

The concentration of B is thus $\dfrac{25.16 \times 0.10}{10.00} = 0.2516\,mol\,dm^{-3}$.

However, since the concentration of A is only known to 2 significant figures, that of B may only be written as $0.25\,mol\,dm^{-3}$.

Errors, rounding up and overtruncation

Most errors in AS coursework will be around 1% leading to three significant figures, e.g. if a concentration of $0.103\,mol\,dm^{-3}$ has an error of 1%, the result lies between 0.102 and 0.104.

There is no harm in putting a fourth digit in, in brackets if unsure.

When the calculator answer gives more than the number of significant figures – say three – the third figure may be rounded up **only** if the fourth figure is 5 or above. Thus 4.268 may be rounded up to 4.27 which is more accurate than 4.26.

A serious error is to destroy information gained in the experiment by over-shortening the result as when a concentration of $0.0946\,mol\,dm^{-3}$ is returned as 0.09 or even 0.1!

Significant figures and decimal places

In calculating the percentage error for the burette above the calculator gives 0.405679513.

What does this mean? The last five numbers mean nothing; since the error is about 0.4%, that is 4 in a thousand, anything after the third or fourth place of decimals has no physical significance.

It is very important that full calculator outputs are not recorded as FINAL answers, only numbers having physical significance should be used. However, the full output may be used at intermediate stages in a calculation.

These are the rules for working out significant figures:

Zeros to left of first non-zero digit are not significant, e.g. 0.0003 has one sig. fig.

Zeros between digits are significant, e.g. 3007 has four sig. figs

Zeros to the right of a decimal point with a number in front are significant,

e.g. 3.0050 has five sig. figs

Decimal places are the number of digits to the right of the decimal point so that 0.044 has 3 decimal places but only two significant figures. Stating the value to two decimal places would give a different value (0.04) than the correct significant figures (0.044).

Always use significant figures and not decimal places when considering errors.

Standard form and ordinary form

Both forms are commonly used, the standard form being useful when large and small numbers are encountered. It is important that the same number of significant figures is used in both, e.g. ordinary form $0.0052\,mol\,dm^{-3}$, standard form $5.2 \times 10^{-3}\,mol\,dm^{-3}$, both have two significant figures.

Unit 1

1.4
Bonding

The usefulness of materials depends on their properties, which in turn depend on their internal structure and bonding. By understanding the relationship between these, chemists can design new useful materials. The types of forces between particles are studied and the importance of electrical attractive and repulsive forces stressed.

You should be able to demonstrate and apply knowledge and understanding of the following:

- ionic bonding in terms of ion formation and the electrical attraction between positive and negative ions.
- Covalent and coordinate bonding as being due to an electron pair with opposed spins between the atoms with each atom providing one electron in covalent bonding and one atom providing both electrons in coordinate bonding.
- Ionic and covalent bonds are extremes and that most bonds are intermediate in character, leading to bond polarity.
- Electronegativity is an approximate way of measuring the electron-attracting power of an atom in a covalent bond and that the greater the difference in electronegativity between the two bond atoms, the more polar will the bond be.
- Metallic bonding is also important and will be dealt with in C1.5.
- Bonding between molecules is much weaker than the bonds within them and is due to permanent and temporary dipoles within the molecule.
- The physical properties of molecules such as boiling temperature are governed by this intermolecular bonding.
- Hydrogen bonding is a stronger intermolecular force occurring when hydrogen is linked between fluorine, oxygen and nitrogen atoms.
- The VSEPR principle that the shapes of molecules can be predicted from the number of bonding and lone electron pairs around the central atom.
- The bond angles arising in linear, trigonal planar, tetrahedral and octahedral molecules and ions.

Content

Bonding

Stretch & Challenge

When atoms bond to form molecules the energy of the bond is released in an exothermic reaction and this process is favoured.

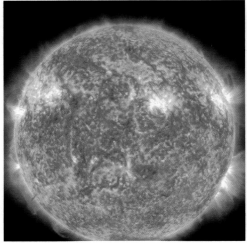

the Sun

▲ 5000° atoms

the Earth

▲ 20° molecules

Apart from the inert gases, elements are rarely found as atoms on Earth since they bond together exothermically to form molecules, and temperatures of above a thousand degrees – as in the Sun – may be necessary to break these bonds. In some cases molecules are formed of a few atoms, as in methane, CH$_4$, In other cases millions of atoms may bond together to form a crystal of salt or diamond or metal. Such structures are called giant molecules.

Key Terms

Ionic bond a bond formed by the electrical attraction between positive and negative ions (cations and anions).

Covalent bond has a pair of electrons with opposed spin shared between two atoms with each atom giving one electron.

Coordinate bond a covalent bond in which both electrons come from one of the atoms.

Chemical bonding

There are different types of **bond**: **ionic**, **covalent** (including **coordinate**) and metallic but the basic cause of bonding is the same in each case: this is that the positively charged nuclei and negatively charged electrons are arranged in such a way that the electrostatic attractions outweigh the repulsions.

Bonds may be shown by 'dot and cross' diagrams in which the outer electrons of one atom forming the bond are shown as dots or open circles, and those of the other atom by crosses.

The electrons may be written as being drawn in circular orbits.

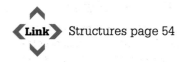

‹Link› Structures page 54

Covalent bonding

Each atom gives one electron to form a bond pair in which the electron spins are opposed.

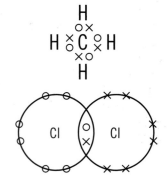

Coordinate bonding

The same as a covalent bond except that both electrons forming the bond pair come from the same atom.

Ionic bonding

One atom gives one or more electrons to the other and the resulting cation (+) and anion (−) attract one another electrically.

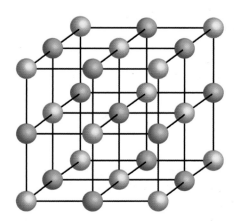

Normally ions form a lattice in which each cation is surrounded by several, e.g. 6, anions and vice versa.

Attractive and repulsive forces

All bonding results from electrical attractions and repulsions between the protons and electrons, with attractions outweighing repulsions.

In covalent bonds the electrons in the pair between the atoms repel one another but this is overcome by their attractions to BOTH nuclei. If atoms get too close together the nuclei and their inner electrons will repel those of the other atom so that the bond has a certain length. Also the electron spins must be opposite for the bonds to form.

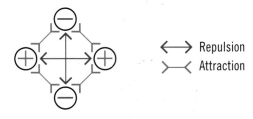

←⟶ Repulsion
⟩—⟨ Attraction

32
Knowledge check

(i) A bond formed between two atoms in which one atom donates an electron completely to the other atom so that both atoms become charged is called a .... bond. A bond in which two atoms each donate an electron to form a bond pair is called a .... bond. A bond in which one atom donates both of the electrons forming the bond is called a .... bond.

(ii) Match each of the substances listed to the type of bond involved:

Bond type; ionic covalent metallic

Molecule KCl Cl_2 Cu CH_4 $CaBr_2$

Link C1.5 page 55

In ionic bonding cations and anions are arranged so that each cation is surrounded by several anions and vice versa to maximise attraction and minimise repulsion. Again repulsions from inner electrons and nuclei prevent the ions from getting too close together.

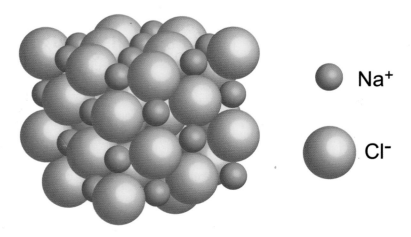

Metallic bonding

This important bond type will be discussed later in Solid structures but essentially consists of a lattice of positive ions held together by a 'sea' of delocalised electrons given up by each atom.

Extra Help

Bond polarity means that although the molecule is neutral overall, one end of the bond has a slightly positive charge and the other end a balancing, slightly negative charge giving a DIPOLE.

Bond polarity is governed by the difference in electronegativity between the two atoms forming the bond. Electronegativity is a measure of the ability of an atom in a covalent bond to attract the electron pair and, on one scale, ranges from 0.7 in caesium to 4.0 in fluorine, with Cs being electropositive and F highly electronegative.

Electronegativity and bond polarity

In a covalent bond the electron pair is not usually shared exactly evenly between the two atoms unless they are the same. Thus one atom will take up a slightly negative charge, the other becoming slightly positive and the bond is now said to be polar. These small charges are written over the atoms using the symbols $\delta+$ and $\delta-$ as shown below:

$$\overset{\delta+}{H} - \overset{\delta-}{F}$$

Coordinate bonds are always polar, since the atom giving both electrons to the bond cannot completely lose its rights over one electron.

Pauling electronegativity values

You may meet other **electronegativity** scales but the one here, devised by Linus Pauling, is the most common.

Key Term

Electronegativity is a measure of the electron-attracting power of an atom in a covalent bond.

Some Pauling electronegativity values:

Li	H	C	N	O	F
1.0	2,1	2.5	3.0	3.5	4.0

Na	Mg	3d metals	S	Cl
0.9	1.2	around 1,6	2.5	3.0

Cs				I
0.7				2.5

Most bonds joining atoms that are not identical will be polar to some extent. At the other extreme, most ionic bonds have some covalent character. However it is simpler for us to treat ionic bonds as completely ionic and take covalent bonds as having varying degrees of polarity. The bond in hydrogen chloride, for example, is about 19% ionic.

Forces between molecules

It is most **important** to distinguish bonding **between** molecules – **intermolecular** – and bonding **within** molecules – **intramolecular**.

Intermolecular bonding is weak and governs physical properties such as boiling temperature; bonding within molecules is strong and governs chemical reactivity. In methane, for example, the forces between the molecules are very weak and the molecules separate, i.e. the liquid boils at minus 162 degrees, but the C—H bonds are very strong and need a temperature of around 600 degrees before they will break. Intermolecular bonding is caused by electrical attraction between opposite charges. Although the molecule may be neutral overall it contains positive and negative charges (electrons and protons) and if the electronegativities of the atoms in the molecule are not the same (see C1.4 page 48), the molecule will have a dipole with parts that are relatively positive and negative in charge. If these dipoles arrange themselves so that the negative region of one molecule is close to the positive region of another molecule, there will be a net attraction between them.

$$\delta^+ \text{——} \delta^- \quad \delta^+ \text{——} \delta^-$$
$$\delta^- \text{——} \delta^+ \quad \delta^- \text{——} \delta^+$$

▲ Permanent partial charge

Even molecules with no dipole show intermolecular bonding, e.g. helium atoms come together to form a liquid at 4 K. This is because the electrons are in constant motion around the nuclei so that the centres of positive and negative charge do not always coincide and give a fluctuating dipole. These come into step with one another as one dipole induces an opposite dipole in a nearby molecule giving an attraction between them.

$$\delta^+\delta^+ \text{——} \delta^-\delta^- \quad \delta^+\delta^+ \text{——} \delta^-\delta^-$$
$$\delta^-\delta^- \text{——} \delta^+\delta^+ \quad \delta^-\delta^- \text{——} \delta^+\delta^+$$

▲ Fluctuating induced charge

To sum up we have here two types of intermolecular bonding, the first dipole–dipole and the second induced dipole–induced dipole and these two together are called **van der Waals forces**.

Strength

Bonding inside molecules is some 100 times stronger than between them with van der Waals strength being around $3 \, kJ \, mol^{-1}$.

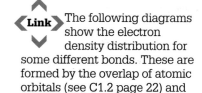

Link The following diagrams show the electron density distribution for some different bonds. These are formed by the overlap of atomic orbitals (see C1.2 page 22) and show the electron density.

Electronegativity Bond difference

3.1	NaF	
0.9	H—Cl	
0.00	H—H	

Key Terms

Intermolecular bonding is the weak bonding holding the molecules together, e.g. as in liquids and governs the physical properties of the substance.

Intramolecular bonding is the strong bonding between the atoms in the molecule and governs its chemistry.

van der Waals forces include all types of intermolecular force whether dipole or induced dipole.

33 Knowledge check

Use the electronegativity values above to decide which of the following molecules is the most polar.

$I_2 \quad O_2 \quad NaCl \quad MgO \quad HF$

34 Knowledge check

State which of the changes A, B, C, D below are in intermolecular bonding and which in intramolecular bonding.

A $Ne(l) \longrightarrow Ne(g)$

B $N_2O_4 \longrightarrow 2NO_2$

C $KI + \frac{1}{2}Cl_2 \longrightarrow KCl + \frac{1}{2}I_2$

D $H_2O(l) \longrightarrow H_2O(g)$

! Extra Help

Inter means between, as in international sport.

Intra means inside, as in intravenous injection.

Hydrogen bonding

This is a special intermolecular bonding force that only occurs between molecules that contain hydrogen atoms bonded to very electronegative elements having lone pairs, namely fluorine, oxygen and nitrogen. Although weak compared with the bonding taking place inside molecules, it is much stronger than ordinary van der Waals forces. Typical strengths for hydrogen bonds are 30 kJ per mol as against 3 kJ for van der Waals and 300 kJ for bonding within molecules. Hydrogen bonding is stronger than van der Waals since the small hydrogen atom is sandwiched between two electronegative elements and allows a close approach.

Hydrogen bonds in action

▲ H bonds in ice

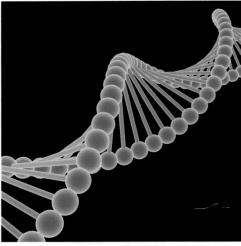

▲ H bonds in DNA

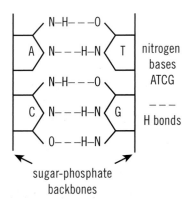

Hydrogen bonding in water and ice

We see that the hydrogen atom is especially δ+ being attached to the electronegative oxygen atom so that the oxygen on the other molecule is closely attracted to it. Also the bonding is strongest when the three atoms are in a straight line and note that the internal O—H bond in the molecule is shorter than the dotted hydrogen connected to the other molecule. Since oxygen has two lone pairs and two hydrogen atoms, a tetrahedral hydrogen-bonded structure is formed.

Effects of hydrogen bonding on boiling temperatures and solubility

Melting and especially boiling temperatures increase with the strength of intermolecular forces. With van der Waals forces there is a steady increase with these with molecular mass and also as dipoles become larger. However, the boiling temperature diagram shows that hydrogen-bonded molecules completely buck the trend compared with methane to SnH_4. In water the molecules freely hydrogen bond with neighbours and these bonds must be largely broken before boiling can occur so that more energy, i.e. a higher temperature, is needed. There are many similar examples.

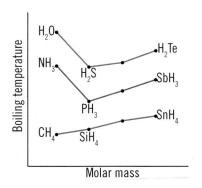

35

Knowledge check

Assign each of the typical energy values below:

A $2{-}4\,kJ\,mol^{-1}$

B $20{-}40\,kJ\,mol^{-1}$

C $200{-}400\,kJ\,mol^{-1}$

to the types of molecular bonding given:

1.   Intramolecular

2.   Intermolecular (hydrogen bond)

3.   Intermolecular (van der Waals).

Solubility in water

As well as hydrogen bonding with other water molecules they themselves can dissolve other molecules such as the lower alcohols with which they can hydrogen bond. On the other hand, non-polar organic molecules such as hydrocarbons that cannot hydrogen bond with water are insoluble in it and prefer to interact with one another through van der Waals forces. The phrase 'like dissolves like' is a useful guide.

Shapes of molecules

The shapes of covalent molecules with more than two atoms and their ions are governed by the electron pairs around the central atom such as C in CH_4. These may be bonding pairs holding the atoms together in covalent bonds or lone pairs on the central atom that are not usually involved in covalent bonding.

There are two points to watch out for:

1. How many electron pairs are connected to other atoms?
2. How many electron pairs in total are around the central atom?

Thus, in ammonia, NH_3, there are three connected pairs and we might first think that the molecule would be planar and propeller-shaped like BF_3. However, the fourth (lone) pair pushes the molecule into an umbrella shape and in fact when H^+ is added to give NH_4^+ we have a tetrahedron. Similarly we might expect water (H_2O) to be linear like BCl_2 but the two lone pairs bend it strongly.

! Extra Help

We could also say that oil and water do not mix. Organic molecules not containing nitrogen, oxygen or fluorine cannot hydrogen bond with water and thus dissolve in it. Conversely, water molecules added to petrol cannot hydrogen bond with it and so bond with one another in separate drops.

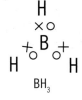

BH_3

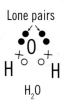

Lone pairs

H_2O

Since all electron pairs repel one another, the molecular shape taken up is that allowing the pairs to keep as far away from each other as possible to minimise the repulsion energy. While bonding pairs are spread out between the two atoms bonded in the molecule, lone pairs stay close to the central atom and so repel more than bonding pairs, giving the repulsion sequence:

lone pair–lone pair > lone pair–bond pair > bond pair–bond pair.

Thus in NH_3 with one lone pair and three bonding pairs the repulsion between the lone pair and the bond pairs is greater than that between the bond pairs themselves so that the H—N—H angle is closed up from 109° to 107°.

These ideas are applied in Valence Shell Electron Pair Repulsion Theory (VSEPR) that follows.

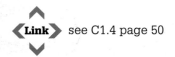
Link see C1.4 page 50

36

Knowledge check

(i) Tests on shapes

Give the shapes of the following molecules

CCl_4 NH_4^+ BeI_2 BCl_3 PF_5

(ii) Match the molecules below with the bond angles listed.

$AlCl_3$ BeH_2 CF_4 SF_6 PF_5

90° 109° 120° 180°

NB Two angles may be present in one case.

YOU SHOULD KNOW ››

››› As stated above, the valence shell is usually the outer electron shell involved in bonding, and commonly contains both electrons and vacancies. Valency is the number of bonds that an atom can make with other atoms. If there are two electrons in the valence shell then the atom can use these to form bonds with two other atoms and is thus said to be divalent.

VSEPR theory

Valence Shell Electron Pair Repulsion (VSEPR) theory lets us predict the shape of simple molecules in which bonded atoms are arranged around a central atom. The valence shell is the electron shell in which bonding occurs. In VSEPR the number of electron pairs is first found to give the general shape of the molecule, since the repelling pairs keep as far away from one another in space as possible. It is worth repeating the repulsion sequence on the shapes of molecules from earlier.

lone pair–lone pair > lone pair–bond pair > bond pair–bond pair and seeing how it applies in the figure on page 51.

The shape follows directly from the number of pairs as below:

No of pairs	Shape	Bond angle	Example
2	linear	180°	$BeCl_2$
3	trigonal planar	120°	BF_3
4	tetrahedral	109.5°	CH_4
5	trigonal bipyramid	90°/120°	PCl_5
6	octahedral	90°	SF_6

Secondly, the exact angles between the bonds will change somewhat depending on the repulsion sequence above. Thus in water with two lone pairs the normal tetrahedral bond angle of around 109° for H—O—H is repelled down to 104° by lone pair – lone pair and lone pair – bond pair repulsion.

You should be able to predict the shape of any simple molecule given its formula using VSEPR. Note that the same rules apply where the covalent molecule is an ion such as NH_4^+ where all electrons are now in bond pairs and the H—N—H bond angles are all 109.5°.

Note also that you are asked:

1. To know and explain the shapes of BF_3, CH_4, NH_4^+ and SF_6.

2. Predict and explain the shapes of other simple species having up to six electron pairs in the valence shell of the central atom.

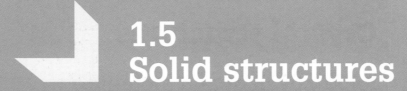

Unit 1

1.5
Solid structures

The properties and usefulness of solid materials depend very much on their structure and bonding at the molecular level. The more that these are understood, the more successful research into design and improvement can be carried out.

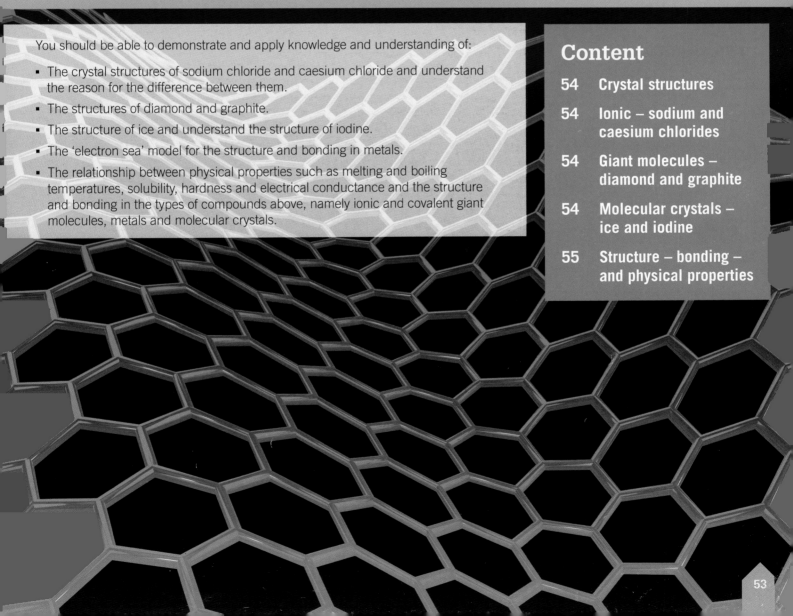

You should be able to demonstrate and apply knowledge and understanding of:

- The crystal structures of sodium chloride and caesium chloride and understand the reason for the difference between them.
- The structures of diamond and graphite.
- The structure of ice and understand the structure of iodine.
- The 'electron sea' model for the structure and bonding in metals.
- The relationship between physical properties such as melting and boiling temperatures, solubility, hardness and electrical conductance and the structure and bonding in the types of compounds above, namely ionic and covalent giant molecules, metals and molecular crystals.

Content

Crystal structures

Ionic, covalent and metallic

You should know how to describe the ionic crystal structures of NaCl and CsCl,

▲ CsCl structure

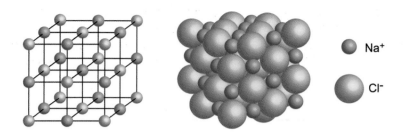

▲ NaCl

Na⁺

Cl⁻

Stretch & Challenge

Draw the iodine structure. In solid iodine it is very important to distinguish between the strong covalent bonds holding iodine atoms together in the I_2 molecule and the weak intermolecular forces that hold the I_2 units in the molecular crystal.

In ionic halides, oppositely charged ions pack around one another in such a way as to increase the bonding energy by maximising electrostatic attraction and minimising repulsion. Thus each cation is surrounded by 6 or 8 anions and vice versa, the actual number depending only on the relative sizes of the anion to cation in the chlorides. Eight chloride anions can fit around the larger Cs ion while the smaller Na can only accommodate six. The **crystal co-ordination numbers** are therefore 8:8 in CsCl and 6:6 in NaCl.

YOU SHOULD KNOW ›››

›››Over three-quarters of the elements are metals.

›››Typically each positive ion in the close-packed lattice is surrounded by eight or twelve others.

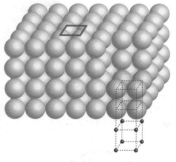

▲ Metal structure

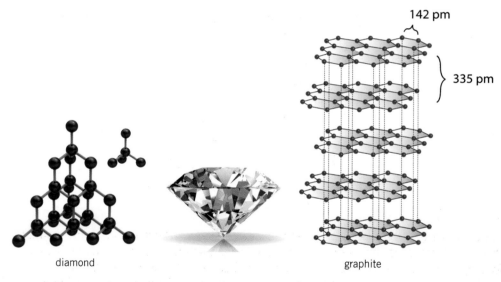

diamond

graphite

142 pm

335 pm

▲ The covalent structures of diamond and graphite

With diamond, the tetrahedral strong covalent arrangement and build-up of a three-dimensional giant structure should be shown while, with graphite, layers made up of covalent hexagons held together by weak forces are needed. You should know and understand the molecular crystal structures of iodine and ice.

Link see C1.4 page 51
Solubility in water

YOU SHOULD KNOW ›››

›››A monolayer is a layer of material that is only one atom or molecule thick; that will be about 10^{-9} metres.

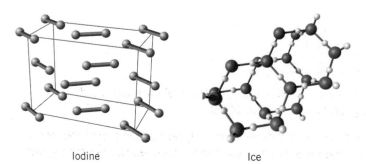

Iodine

Ice

In solid iodine it is very important to distinguish between the strong covalent bonds holding the iodine atoms together in the I_2 molecule and the weak intermolecular forces that hold the I_2 units together in the molecular crystal.

In ice, hydrogen bonds between the water molecules hold them together in a fairly open tetrahedral structure while the strong bonds within the water molecules are polar covalent.

Metals

While the actual structures of metals are not required, you need to understand the general concept that atoms of metallic elements each donate one or more electrons to form a delocalised electron sea or gas. This sea surrounds the close-packed positive ions so formed and binds them together through the attraction between opposite charges.

Structure and physical properties

It is important to be able to explain the properties of all the solid types previously covered in terms of their structures.

The giant ionic structures such as the chlorides are generally hard, brittle and high-melting due to the strong ionic bonds. There is no electrical conduction in the solid state since the ions are fixed in the crystal but the molten salts and aqueous solutions of them do conduct since they are now free to move when a voltage is applied. Ionic solids may or may not be soluble in water, depending on energetics or chemical reaction factors, but most ionic chlorides are soluble.

Here the polar water molecules orient themselves around the anion and cation as shown to give a hydration energy term that favours solubility.

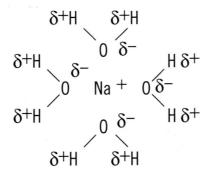

The covalent giants, diamond and graphite, are very high-melting and insoluble in water; diamond is very hard, with each carbon atom being covalently bonded to four others and forming a three-dimensional structure in space, but the weak layer structure in graphite renders it softer and useful as a lubricant. Also graphite conducts electricity owing to the π electron delocalisation in the ring plane while diamond and iodine do not. Iodine is soft and volatile since the I_2 units are held together only by weak van der Waals forces.

Electron delocalisation in metals gives good electrical and thermal conductivity but their melting temperatures and hardness increase with the number of electrons per atom involved in bonding, e.g.

Metal	Na	Ca	V
no of bonding electrons	1	2	5
melting temp. °C	98	850	1900

Sodium is a soft metal that is easily cut with a knife but vanadium is very hard. Also remember mercury, melting temperature minus 39° C!

In practice the actual properties of a solid depend not only on the bonding at the atomic level but on the way in which the units, such as the carbon nanotubes, are held together.

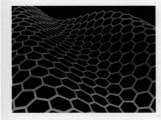
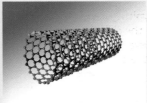

⩾0.5 Non Polar Covalent   0.5-1.9 Polar   1.95 ionic

Unit 1

1.6
The periodic table

The periodic table is an essential tool in dealing with the vast number of chemical elements and compounds that exist enabling us to see trends and make predictions. In this topic we see how the table is set up and operates and study the important chemistry of the s-block Group 1 and 2 elements and compounds and the halogens in Group 7.

Content

You should be able to demonstrate and apply knowledge and understanding of:

- The arrangement of elements in the table in terms of the electronic structures in the s, p and d blocks.
- Oxidation and reduction in terms of electron transfer.
- The trends in ionisation energy, electronegativity and physical properties across periods and down groups.
- The common reactions of the s-block elements with water, acids and oxygen and their flame colours.
- The common reactions of Group 2 ions with hydroxide, carbonate and sulfate ions and the thermal stability and solubility trends of the common salts.
- The reaction of the halogens with metals and the trend in their volatility.
- Their relative oxidising power and understand the redox displacement reactions.
- The reaction of halide ions with Ag^+ followed by dilute ammonia.
- The use of chlorine and fluoride ions in water sterilisation.
- How to perform simple gravimetric analysis and salt formation and crystallisation.
- How to identify unknown solutions by logic and qualitative analysis.

Basic structure

An understanding of the general trends and behaviour gives us great predictive power. The chemistry of the elements is governed largely by their outer electrons so that arranging elements in groups according to their outer structure simplifies our study of their behaviour. Ionisation energy (I.E.) and electronegativity (x) increase diagonally across the table (i.e. across a period and up a group). For example, the the I.E.s for Cs and F are 376 and 1680 kJ respectively.

Electrons are thus readily lost in the s block giving cations and ionic compounds; entering the p block in Group 3 I.E.s become too high so that electron sharing (covalency) is usual, but the more electronegative elements in Groups 6 and 7 can accept electrons to form anions in ionic compounds.

Valency normally rises with group number up to a maximum of four and then falls (8 minus the group number) to one in Group 7.

Elements are generally metals when IEs are low in the left and lower regions of the table and the d block transition elements; non-metals occur in the high I.E., upper right, portion and semiconductor elements, such as silicon, are found between these two regions.

Melting temperature trends are more complex, depending on atomic mass, bond type and the type of solid structure but decrease down Group 1, rise down Group 7, increase across a period up to Group 4 (carbon melts above 3500°) and the drop sharply as elements form diatomic molecules held in the solid by weak intermolecular forces.

	s block		d block	p block					
Group	1	2	transition metals	3	4	5	6	7	8 inert gas
Oxidn no	1	2					−2	−1	
Redox	reducing			oxidising					
Ions	cations			anions					
Oxides	basic			acidic					
MT	decrease down		increase to Group 4	inc. down					
Element type	metals			non-metals					

I.E. and x increase diagonally across and up

▲ Trends in the periodic table

HOW SCIENCE WORKS

The development of scientific knowledge and understanding of chemistry over many years in the 19th century led to the periodic table of Mendeleev that has given a great increase in understanding and predictive power in the subject. Gaps left in the table directed the search to then–unknown elements, such as the important semiconductor germanium, which when found had exactly the properties predicted by Mendeleev. The regular trends in properties down and across the groups led to a great simplification in the chemistry of the elements and the table served to guide the development of the electronic structure advances at the beginning of the 20th century.

Stretch & Challenge

Salt solubility

Salt solubility is quite a complex matter but the strength of the lattice in an ionic crystal is important. In $MgSO_4$, for example, the Mg ion is very much smaller than the SO_4 so that the ions cannot fit together well to form a strong lattice and $MgSO_4$ is soluble. In $BaSO_4$ the Ba ion is much bigger than Mg, the ions fit together well in the lattice and $BaSO_4$ is insoluble in water.

37

Knowledge check

(i) In the periodic table the values of ionisation energy and electronegativity .... down a group and .... across a period.

(ii) The solubility of Group II hydroxides .... down the group and the solubility of Group II sulfates .... down the group.

Group 1	all salts soluble
Group 2	
$M(OH)_2$	solubility increases down group – $Mg(OH)_2$ insoluble
MSO_4	solubility decreases down group – $BaSO_4$ insoluble
MCO_3	all insoluble

▲ Solubility in water

38

Knowledge check

Redox (oxidation/reduction) reactions involve the transfer of .... from one agent to another. A substance that gives one .... to another substance is called a ·········· agent and is thus .... by this process while the receiving agent is said to be ....

39

Knowledge check

(i) Give the oxidation states of all the atoms in the following compounds:

$FeCl_3$ H_2O_2 $K_2Cr_2O_7$
Br_2 H_2CO_3

(ii) Write down the oxidation states of **all** the atoms on both sides of the following unbalanced reaction:

$NaI + H_2SO_4 \longrightarrow Na_2SO_4 + I_2 + H_2S + H_2O$

◀ **Link** ▶ C1.1 see page 12

▲ Sodium metal on water

Redox

Many chemical reactions involve the loss or gain of electrons, a species being oxidised if it loses electrons and reduced if it gains them. Since electrons do not vanish or appear from nowhere, all these reactions involve a transfer of electrons from the species being oxidised to the one being reduced.

E.g., in $Na + \frac{1}{2}Cl_2 \longrightarrow Na^+ + Cl^-$, the Na is being oxidised and loses an electron and the Cl gains an electron and is reduced.

The popular nemonic OILRIG is helpful if used carefully with the atom saying '**O**xidised **I** **L**ose electrons, **R**educed **I** **G**ain electrons.

Confusion is very common.

Oxidation numbers (states)

This is a useful accounting system for **redox** with simple rules:

1. All elements have an oxidation state of zero.

2. Hydrogen in compounds is usually 1 (or +1).

3. Oxygen is usually −2 or −II.

4. Group 1 and 2 elements in compounds are 1 and 2 respectively.

5. Group 6 and 7 elements in compounds are usually −2 and −1 respectively.

6. An element bonded to itself is still 0.

7. The oxidation numbers of the elements in a compound or ion must add up to zero or the charge on the ion.

Important
The oxidation number does not imply a charge, e.g. in MnO_4^- the oxidation numbers are Mn(7) and four O(4×-2) giving an overall charge of minus 1. The Mn is **not** 7+.

The S-block elements

The elements are all reactive, electropositive (low electronegativity) metals forming cations with oxidation numbers 1 or 2 respectively.

Oxides are formed with oxygen/air as in $Ca + \frac{1}{2}O_2 \longrightarrow CaO$. Hydrogen is liberated with water and an oxide or hydroxide formed;

$$Na + H_2O \longrightarrow NaOH + \tfrac{1}{2}H_2$$

The reaction of Group 1 and Group 2 elements with acids is similar to this except that a salt is formed as in

$$Mg + 2HCl \longrightarrow MgCl_2 + H_2$$

and the elements in both groups show their typical reaction as reducing agents, donating electron(s) to reduce the acid or water to hydrogen and being themselves oxidised.

$$Mg + 2HCl \longrightarrow Mg^{2+} + H_2 + 2Cl^-$$
Oxidation number $\quad$ O $\quad$ 2(1)(-1) $\quad$ 2 $\quad$ O $\quad$ 2(-1)

In all these cases reactivity increases down the group and Group 1 elements are more reactive than Group 2.

Lithium reacts slowly with water while potassium reacts violently; magnesium react slowly but barium is faster. All the s-block metals react vigorously with acids.

Flame test → Li → Red (Scarlet)   K → (lilac)   Ca → brick red
Na → Yellow   Ba → Green
              Sr → Crimson red   Mg → none

All react with oxygen and burn in air while caesium inflames spontaneously.

The oxides are all basic, i.e. they react with acids to give salts as in

$$CaO + 2HCl \longrightarrow CaCl_2 + H_2O$$

Remember that the Group 2 hydroxide formulae are $M(OH)_2$ since (OH) is −1, i.e. [O(−2) H(+1)].

While Group 1 salts are all soluble the reactions of Group 2 ions with hydroxide, carbonate and sulfate ions give a variety of results that must be known. $Mg(OH)_2$ is insoluble in water but solubility increases down the group; $BaSO4$ is insoluble and solubility increases up the group. Although you are not required to remember any numerical values, they should help understanding of the scale of changes. Hydroxide solubilities rise from 0.01 g dm^{-3} with Mg to 3.9 g dm^{-3} in Ba; sulfate solubilities fall down the group from 330 g dm^{-3} in Mg to 0.002 g dm^{-3} in Ba.

All Group 2 carbonates are insoluble and all s-block nitrates soluble.

The thermal stability of Group 2 hydroxides and carbonates increases down the group; for example magnesium carbonate decomposes at 400° as against 1300° for barium carbonate.

Although not in the Specification, it is useful in practical work to know that all the nitrates and chlorides of s-block elements are soluble in water.

Flame colours: All of the common elements of Groups 1 and 2 except Mg show characteristic flame colours that must be known and that are useful in qualitative analysis.

You should be aware of the great importance of calcium carbonate in both living and inorganic systems and of calcium phosphate minerals in living bones and skeletons. Calcium and magnesium ions play a vital role in the biochemistry of living systems – chlorophyll, muscle operation, etc., and the carbonates exist in huge amounts in rocks – such as chalk, limestone and dolomite.

Utalisation d ATP

The halogens

These reactive, **electronegative elements** typically form anions having an oxidation state of −1 so that oxidation is the usual reaction as in:

$$Na(0) + \tfrac{1}{2}Cl_2(0) \longrightarrow Na^+(1) + Cl^-(-1)$$

with the Na being oxidised and the oxidising chlorine being reduced from 0 to −1. The tendency to form anions decreases down the group from fluorine to iodine with fluorine being the most electronegative element.

The melting temperatures of the elements increase down the group from gaseous fluorine to solid iodine owing to the increasing intermolecular forces holding the diatomic elements together in a liquid or solid. This increase is due to the increasing number of electrons in the molecules contributing to the induced dipole–induced dipole intermolecular force.

The halogens react with most metals to form halides with the reactivity decreasing down the group from fluorine to iodine. A similar feature is shown in displacement reactions in which a halogen higher in the group displaces one lower in the group from a salt as in:

$$Cl_2 + 2NaBr \longrightarrow Br_2 + 2NaCl$$

This essentially reflects the decrease in oxidising power down the group with chlorine oxidising the bromide ion to bromine and being itself reduced to chloride:

$$Cl_2(0) + 2 Br^-(-1) \longrightarrow 2Cl^-(-1) + Br_2(0)$$

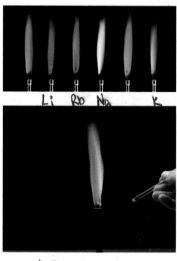

▲ Flame tests

▲ $CaCO_3$ fossils

▲ Common silver halide
precipitates

The reaction of halide ions with silver ions in dilute nitric acid is important in qualitative analysis in both organic and inorganic chemistry. The general reaction is:

$$Ag^+(aq) + X^-(aq) \longrightarrow AgX(s)$$

The precipitate colours are chloride (white), bromide (pale cream) and iodide (pale yellow) and only the silver chloride dissolves in dilute ammonia. This gives a simple way of identifying the halogen present.

Uses of chlorine and fluoride in water treatment

Chlorine is widely used in water treatment and makes water safe to drink and use by killing pathogenic bacteria and viruses and preventing the outbreak of serious diseases such as typhoid and cholera. It is commonly added as the gaseous element and sets up the equilibrium below in water:

$$Cl_2 + H_2O \longrightarrow HOCl + HCl$$

There appear to be only beneficial effects below one part per million (ppm).

Fluoride is generally added to water and to toothpaste and again appears to have only beneficial effects below 1 ppm, both on reducing tooth decay caused by cavity formation and in strengthening bones, thus reducing osteoporosis. It is added as fluorosilicic acid, sodium fluorosilicate or sodium fluoride.

Although effective, there has been opposition to its addition to public water supplies on ethical grounds. However, the gains outweigh the losses, especially in the case of chlorine, where the once deadly disease of cholera has been largely eliminated.

42

Knowledge check

Name the reactants required to form copper(II) chloride, water and carbon dioxide.

Practical work

Soluble salt formation

For example, copper(II) sulfate can be formed by neutralising sulfuric acid with the insoluble base copper(II) oxide,

$$H_2SO_4(aq) + CuO(s) \longrightarrow CuSO_4(aq) + H_2O(l)$$

These are the steps in the formation of the salt.

1. Some copper(II) oxide is added to dilute sulfuric acid. More is added until no more dissolves. (Warming might be necessary.) The solution turns blue.
2. All the acid has been used up. The excess solid is removed by filtering. This leaves a blue solution of copper(II) sulfate in water.
3. The solution is heated to evaporate some of the water.
4. It is left to cool. Blue crystals of copper(II) sulfate start to form.

The water is not fully evaporated because if this happens, a powder will form rather than crystals.

If copper(II) carbonate is used the method is exactly the same but effervescence (fizzing) is seen when the carbonate is added to the acid because carbon dioxide is given off. When no more effervescence is seen, all the acid has been used up.

▲ Adding CuO

▲ Forming CuSO₄

▲ Evaporating CuSO₄

▲ CuSO₄ crystals

Gravimetric analysis

Aim
To measure the exact concentration of calcium ions in a calcium chloride solution by precipitation as calcium carbonate and weighing.

Procedure
Place **exactly** 100 cm³ of an approximately 0.2 mol dm⁻³ calcium chloride solution into a clean beaker and stir in around 110 cm³ of an approximately 0.2 mol dm⁻³ solution of sodium carbonate. Allow the precipitate to settle and check for complete precipitation by adding more drops of carbonate to the clear solution.

When no further precipitate forms, filter off the precipitate making sure that all of it is transferred to the filter.*

Wash the precipitate with distilled water to remove any impurities and dry in a drying oven. Weigh the dry precipitate plus filter and subtract the weight of the filter to give the mass of the calcium carbonate. Use this to calculate the mass of calcium in the 100 cm³ sample and thus the exact concentration of the calcium chloride solution.

*NB The filtration procedure will depend on the apparatus available. Most simply vacuum filtration though a sintered glass crucible is used but otherwise filter paper in a filter funnel is employed.

Calculation of concentration
Volume of Ca solution = V dm³

Mass of precipitate/g = m (= mass of ppt + filter − mass filter)

Mass of $CaCO_3$/dm³ = m/V

Concn of calcium ions = m/VM where M is molar mass of $CaCO_3$.

The numbers of calcium ions in the carbonate and chloride solution are of course identical.

PRACTICAL CHECK

Gravimetric analysis and Qualitative analysis are **specified practical tasks**

Qualitative analysis

See questions 1,6,7 and 9 in Exam Practice Questions

1.7
Simple equilibria and acid–base reactions

So far, in this book, chemical reactions have been written as reactants $\longrightarrow$ products. However, chemical reactions do not only move in the forward direction from left to right in a chemical equation. Some reactions move in the backward direction, from right to left. This unit studies the relationship between the forward and backward reactions and their overall effect on the yield of reaction.

The word acid comes from the Latin 'acerbus' meaning sour. However, not all sour-tasting substances are acids, and very few acids are safe to taste to see if they are sour! In this unit you will learn about acids, their reactions, how to measure their acidity and how to neutralise them.

Content

You should be able to demonstrate and apply knowledge and understanding of:

- What is meant by a reversible reaction and dynamic equilibrium.
- Le Chatelier's principle in deducing the effects of changes in temperature, concentration and pressure.
- The equilibrium constant (K_c) and calculations involving given equilibrium concentrations.
- Acids as donors of $H^+(aq)$ and bases as acceptors of $H^+(aq)$.
- The relationship between pH and $H^+(aq)$ ion concentration $pH = -\log[H^+(aq)]$.
- Acid–base titrations.
- The difference between strong acids and weak acids in terms of relative dissociation.

Reversible reactions

Not all chemical reactions 'go to completion', i.e. the reactants change completely to form products. Reactions do not only move in the forward direction, some reactions also move in the backward direction and products change back into reactants. These reactions are called **reversible** and the symbol '$\rightleftharpoons$' is used in equations.

You will already be familiar with a number of reversible reactions at home, in the laboratory and in industry. If you lower the temperature of water below 0 °C it freezes to ice. If the ice is allowed to reach room temperature it melts back to water. The process can be represented as:

$$\text{Water} \rightleftharpoons \text{Ice} \quad \text{or} \quad H_2O(l) \rightleftharpoons H_2O(s)$$

Blue copper(II) sulfate crystals have the formula $CuSO_4.5H_2O$ (the water is called water of crystallisation). When the copper(II) sulfate is heated, the water of crystallisation is given off as steam, leaving a white powder called anhydrous copper(II) sulfate. When water is added to the white powder, the powder gets hot and turns blue. The process can be represented as:

$$\underset{\text{blue}}{CuSO_4.5H_2O} \rightleftharpoons \underset{\text{white}}{CuSO_4} + 5H_2O$$

The Haber process, for the formation of ammonia from nitrogen and hydrogen, is a well-known industrial example:

$$N_2(g) + 3H_2(g) \rightleftharpoons 2NH_3(g)$$

YOU SHOULD ›››

››› understand what is meant by a reversible reaction and dynamic equilibrium

Key Terms

A reversible reaction is one that can go in either direction depending on the conditions.

Dynamic equilibrium is when the forward and reverse reactions occur at the same rate.

Dynamic equilibria

Equilibrium is a term used to denote balance. The two main types encountered in everyday life are static and **dynamic equilibrium**. Imagine you are looking at a classroom in school. All of the class's 30 seats are being used. The students are sitting a test and half an hour later, you see that the same pupils are sitting at the same tables, so the situation has not changed, i.e. it is static. You go into another classroom, again all of the class's 30 seats are being used, but this time the students are doing group work. After a while some of the students move groups but all the seats are still occupied. This situation continues for the rest of the lesson. There is a balance between the number of students leaving a table and arriving. Although the number of pupils has not changed there is constant motion, i.e. it is a dynamic equilibrium.

An example of dynamic equilibrium is dissolving an ionic compound in water. When copper(II) sulfate crystals are added to water, the crystals begin to dissolve and the solution turns blue. The more copper(II) sulfate added, the deeper the blue colour. When no more dissolves and copper(II) sulfate crystals remain in the solution, the solution has become saturated and the intensity of the blue colour remains constant.

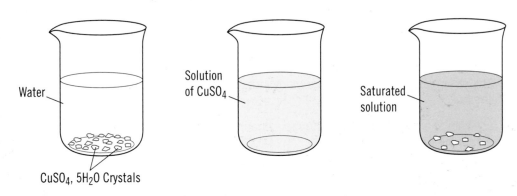

Water

Solution of $CuSO_4$

Saturated solution

$CuSO_4$, $5H_2O$ Crystals

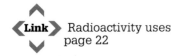

Link Radioactivity uses page 22

At this point the solution is in equilibrium with the undissolved solid. Although nothing seems to be happening since the concentration of the saturated solution remains the same, the copper(II) sulfate is still dissolving but as it does so copper(II) sulfate is recrystallising from solution at the same rate.

This can be proved by using radioactive tracers. If crystals containing the radioactive ^{35}S are added to the saturated solution, after a time it is seen that radioactivity is divided between the solution and the undissolved crystals.

Dynamic equilibria can also develop during chemical changes. The balance of reactants and products at equilibrium is called the equilibrium mixture. If the equilibrium mixture contains mostly products and hardly any reactants, we say that the reaction has gone to completion, (e.g. the burning of magnesium in air). On the other hand, if the equilibrium mixture consists mostly of reactants and hardly any products, we say that the reaction does not happen under these conditions, (e.g. in water vapour at room temperature, it is difficult to detect any hydrogen or oxygen present).

At equilibrium there is no observable change; the properties that we can see or measure (e.g. concentrations of reactants and products) remain constant. These are called macroscopic properties. However, the system is in constant motion – the dynamic change happening at a molecular level. For this to happen, the reactants and products must be in contact at all times so a closed system is needed, i.e. one in which substances cannot leave and cannot enter.

When equilibrium is established in a chemical process, normally the concentrations of the reactants and products change rapidly at first and then reach steady values. At this point, the reactants are being converted into products at exactly the same rate as products are being converted into reactants.

▼ **Study point**

Features of an equilibrium are:

- It is dynamic at a molecular level.
- The forward and reverse reactions occur at the same rate.
- There is a closed system.
- Macroscopic properties remain constant.

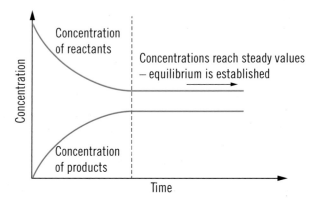

YOU SHOULD ›››

››› understand Le Chatelier's principle and how it is used in deducing the effect of changes in temperature, concentration and pressure on a system in equilibrium

Key Terms

Position of equilibrium is the proportion of products to reactants in an equilibrium mixture.

Le Chatelier's principle states that if a system at equilibrium is subjected to a change, the equilibrium tends to shift so as to minimise the effect of the change.

Position of equilibrium

If we change the conditions of a chemical reaction in equilibrium, the equilibrium is also changed. For example, when a change in the concentration of one substance is made, the concentration of all the other substances involved in the equilibrium will change, i.e. the concentration of the reactants and products will change. The proportion of products to reactants in an equilibrium mixture is known as the **position of equilibrium**.

The response of equilibrium systems to changes of concentration, pressure and temperature were observed by Henri Le Chatelier and in 1884 he presented a principle which has important implications for many industrial processes. **Le Chatelier's principle** states that:

If a system at equilibrium is subjected to a change then the position of equilibrium will shift to minimise that change.

Effect of concentration change

Dissolving hydrated copper(II) sulfate crystals in water gives a blue solution because the ion $[Cu(H_2O)_6]^{2+}$ is formed. When concentrated hydrochloric acid is added to a solution of copper(II) sulfate the following equilibrium is established:

$$[Cu(H_2O)_6]^{2+}(aq) + 4Cl^-(aq) \rightleftharpoons [CuCl_4]^{2-}(aq) + 6H_2O(l)$$
pale blue   yellow–green

Adding more concentrated hydrochloric acid to the solution adds chloride ions, so the system will try to minimise this effect by decreasing the concentration of chloride ions and so the position of equilibrium will move to the right and form more $CuCl_4^{2-}$ ions making the solution yellow–green.

In the same way addition of water to the reaction mixture moves the position of equilibrium to the left making the solution blue.

Effect of pressure change

Pressure has virtually no effect on the chemistry of solids and liquids. However, it has significant effects on the chemistry of reacting gases.

The pressure of a gas depends on the number of molecules in a given volume of gas. The greater the number of molecules, the greater the number of collisions per unit time, therefore the greater the pressure of the gas.

Nitrogen dioxide, NO_2, a red-brown gas, is a major atmospheric pollutant from car exhausts. When two molecules of nitrogen dioxide join together to form one molecule of the colourless dinitrogen tetroxide, N_2O_4, the following equilibrium exists:

$$2NO_2(g) \rightleftharpoons N_2O_4(g)$$
brown   colourless

Since there are two moles of gas on the L.H.S. and one mole of gas on the R.H.S., the L.H.S. is the side at the higher pressure. If the total pressure is increased, the equilibrium will shift to minimise the increase. The pressure will decrease if the equilibrium system contains fewer gas molecules. Therefore the position of equilibrium moves to the right, increases the yield of N_2O_4 and the colour becomes lighter.

Conversely, reducing the pressure shifts the position of equilibrium to the left and the colour becomes darker.

Effect of temperature change

An endothermic reaction absorbs heat from the surroundings, whereas an exothermic reaction gives out heat to the surroundings. For a reversible reaction, if the forward direction is exothermic, the backward direction is endothermic and vice versa. We can identify if a reaction is exothermic or endothermic by looking at the value of the enthalpy change, ΔH. If ΔH has a negative value the reaction is exothermic, if ΔH is positive the reaction is endothermic. The enthalpy change of the forward reaction will have the same magnitude as, but the opposite sign to the backward reaction.

Again, consider the equilibrium

$$2NO_2(g) \rightleftharpoons N_2O_4(g) \qquad\qquad \Delta H = -24\,kJ\,mol^{-1}$$
brown   colourless

Since the enthalpy change is negative, the forward reaction is exothermic. If the temperature is increased the system will try and minimise this increase. The system opposes the change by taking in heat so the position of equilibrium moves in the

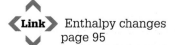

endothermic direction. Therefore the equilibrium moves to the left, decreasing the yield of N_2O_4 and increasing the yield of NO_2 making the mixture appear brown.

In the same way, decreasing the temperature shifts the equilibrium to the right favouring the exothermic direction, increasing the yield of N_2O_4 so the mixture appears a much lighter colour.

Effect of catalysts

A catalyst speeds up a chemical reaction by lowering the activation energy of the reaction. In a reversible reaction a catalyst will increase the rate of the forward and backward reaction to the same extent. Therefore a catalyst does not affect the position of equilibrium, but equilibrium is reached faster.

Equilibrium constant

As we have seen in this chapter, in a reversible reaction the position of equilibrium may lie to the right, i.e. more products; or to the left, i.e. more reactants. However, the equilibrium position changes when the temperature, pressure and concentrations change. The equilibrium position may be described in precise terms by combining the equilibrium concentrations to give a value for an **equilibrium constant**. It is given the symbol K_c where the subscript $_c$ indicates that it is a ratio of concentrations.

An example is the reaction between ethanoic acid and ethanol

$$CH_3CO_2H(aq) + C_2H_5OH(aq) \rightleftharpoons CH_3CO_2C_2H_5(aq) + H_2O(l)$$

The equilibrium constant K_c is given by:

$$K_c = \frac{[CH_3CO_2C_2H_5][H_2O]}{[CH_3CO_2H][C_2H_5OH]}$$

Where $[CH_3CO_2H]$ is the equilibrium concentration of ethanoic acid in $mol\,dm^{-3}$.

In general for an equilibrium: $\quad aA + bB \rightleftharpoons cC + dD$

$$K_c = \frac{[C]^c\,[D]^d}{[A]^a\,[B]^b} \quad \text{where [C] represents the concentration of C at equilibrium, in } mol\,dm^{-3}.$$

Note that:

- The products are put in the numerator (top line) and the reactants in the denominator (bottom line).
- The concentrations are raised to powers corresponding to the mole ratio in the equation.
- The unit of K_c can vary, it depends on the equilibrium.

In the above example:

$$K_c = \frac{[CH_3CO_2C_2H_5][H_2O]}{[CH_3CO_2H][\,C_2H_5OH]}$$

and the units are: $\dfrac{mol\,dm^{-3} \times mol\,dm^{-3}}{mol\,dm^{-3} \times mol\,dm^{-3}}$

therefore the units 'cancel out' and K_c has no units.

However for the equilibrium

$$2NO_2(g) \rightleftharpoons N_2O_4(g)$$

$$K_c = \frac{[N_2O_4]}{[NO_2]^2}$$

and the units are: $\dfrac{mol\,dm^{-3}}{mol\,dm^{-3} \times mol\,dm^{-3}} = dm^3\,mol^{-1}$

A large numerical value of K_c shows that there are more products than reactants in the equilibrium mixture, i.e. the position of equilibrium lies to the right.

A value of less than 1 for K_c shows that there are more reactants than products in the equilibrium mixture, i.e. the position of equilibrium lies to the left.

The value of K_c is constant for a particular equilibrium reaction at a constant temperature. Therefore only a change in temperature can change the value of K_c. Although changing the concentration or pressure can cause a shift in the position of equilibrium it does not change the value of K_c.

K_c has been shown to be constant for a particular reaction by setting up several experiments in which the initial amounts of reactants are varied and the equilibrium concentration of reactants and products are measured (at constant temperature).

Worked example

For the system:

$$2H_2S(g) \rightleftharpoons 2H_2(g) + S_2(g)$$

equilibrium was reached at a temperature of 1400 K.

The equilibrium mixture contained the following concentrations:

$[H_2S] = 4.84 \times 10^{-3}\,mol\,dm^{-3}$, $[H_2] = 1.51 \times 10^{-3}\,mol\,dm^{-3}$, $[S_2] = 2.33 \times 10^{-3}\,mol\,dm^{-3}$.

Calculate the value of the equilibrium constant, K_c, at this temperature and give its units.

$$K_c = \frac{[H_2]^2[S_2]}{[H_2S]^2}$$

$$K_c = \frac{(1.51 \times 10^{-3})^2 \times (2.33 \times 10^{-3})}{(4.84 \times 10^{-3})^2}$$

$$K_c = 2.27 \times 10^{-4}\,mol\,dm^{-3}$$

Knowledge check

Hydrogen and iodine are reacted together to form hydrogen iodide and allowed to reach equilibrium. The equilibrium concentrations are as follows
$[H_2] = 0.11\,mol\,dm^{-3}$,
$[I_2] = 0.11\,mol\,dm^{-3}$,
$[HI] = 0.78\,mol\,dm^{-3}$.
Calculate the value of the equilibrium constant K_c.

Stretch & Challenge

The results for three experiments (carried out at constant temperature) for ammonia synthesis

$$N_2(g) + 3H_2(g) \rightleftharpoons 2NH_3(g)$$

are given in the table

Exp.	Initial concentration (mol dm^{-3})			Equilibrium concentration (mol dm^{-3})		
	N$_2$	H$_2$	NH$_3$	N$_2$	H$_2$	NH$_3$
1	1	1	0	0.922	0.763	0.157
2	0	0	1	0.399	1.197	0.203
3	2	1	3	2.59	2.77	1.82

Show that the value of the equilibrium constant, K_c is unaffected by changes in concentration.

Acids and bases

Acids and **bases** make up some of the most familiar chemicals in our everyday lives, as well as being some of the most important chemicals in laboratories and industries. For these chemicals to act alike there must be some common properties in the chemicals in each of these groups.

Key Terms

An **acid** is a proton (H$^+$) donor.

A **base** is a proton (H$^+$) acceptor.

▼ **Study point**

Bases are normally metal oxides and hydroxides.

An alkali is a soluble base.

Stretch & Challenge

The hydrogen ion, H^+, is simply a proton. Since it has an extremely small diameter (10^{-15} m) compared with other cations (around 10^{-10} m) it has a very large charge density. In aqueous solution it attracts a lone pair of electrons on a neighbouring water molecule to form a co-ordinate bond. The aqueous hydrogen ion, $H^+(aq)$ actually exists as the H_3O^+ ion, called the oxonium ion. It has a roughly pyramidal shape.

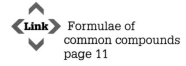
Link Formulae of common compounds page 11

The ion common to all acids is the hydrogen ion, H^+. An acid is a compound that donates H^+ ions (protons) in aqueous solution. This process is called dissociating. Some common acids and their formulae are:

Hydrochloric acid	HCl
Sulfuric acid	H_2SO_4
Nitric acid	HNO_3
Ethanoic acid	CH_3COOH

The equation for hydrochloric acid dissociating is given by:

$$HCl(g) \xrightarrow{\text{water}} H^+(aq) + Cl^-(aq)$$

A base is a compound that accepts H^+ ions from an acid. Some common bases and their formulae are:

Magnesium oxide	MgO	Calcium oxide	CaO
Sodium hydroxide	NaOH	Ammonia	NH_3

If the base dissolves in water it is called an **alkali**. The ion common to all alkalis is the hydroxide ion, OH^-.

$$NaOH(s) \xrightarrow{\text{water}} Na^+(aq) + OH^-(aq)$$

Strong and weak acids

We know that different acids have different strengths. It is possible to taste an acid like citric acid quite safely but sulfuric acid will quickly damage your tongue!

Since acids donate H^+ ions in aqueous solution, the more easily an acid can donate H^+ the stronger the acid.

The general equation for dissociation of an acid is given by:

$$HA(aq) \rightleftharpoons H^+(aq) + A^-(aq) \qquad \text{(A^- represents an anion)}$$

For HCl the equilibrium lies far to the right so the equation is written as:

$$HCl(aq) \longrightarrow H^+(aq) + Cl^-(aq)$$

The acid is fully dissociated or ionised and it is described as a strong acid.

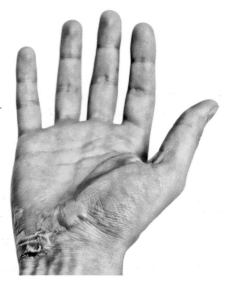

▲ Concentrated acids can burn

Therefore, a **strong acid** is fully dissociated in aqueous solution. The aqueous hydrogen ion concentration is equal in magnitude to the concentration of the acid.

Many acids are far from fully dissociated in aqueous solution and these are described as weak acids, e.g. for ethanoic acid, the equilibrium

$$CH_3CO_2H(aq) \rightleftharpoons CH_3CO_2^-(aq) + H^+(aq)$$

lies to the left. In fact only about four in every thousand ethanoic acid molecules are dissociated into ions.

A **weak acid** is only partially dissociated in aqueous solution. The aqueous hydrogen ion concentration is smaller in magnitude than the concentration of the acid.

The words strong and weak only refer to the extent of dissociation and not in any way to concentration. A **concentrated acid** consists of a large quantity of acid and a small quantity of water. A **dilute acid** contains a large quantity of water.

So it is possible to have a dilute solution of a strong acid e.g. HCl of concentration $0.0001 \, mol \, dm^{-3}$ or a concentrated solution of a weak acid e.g. CH_3CO_2H of concentration $8 \, mol \, dm^{-3}$.

Similarly bases can be classified as strong or weak. An example of a strong base is NaOH and an example of a weak base is NH_3.

▲ Dilute acids can be used to clean glass

The pH scale

Since acids are compounds that donate H^+ ions (aq), the acidity of a solution is a measure of the concentration of the aqueous hydrogen ion, H^+(aq).

However, H^+(aq) concentration varies over a wide range and can be extremely small, e.g. from 1.10^{-14} to $1 \, mol \, dm^{-3}$. This wide variation made it very difficult for people to deal with the concept of acidity.

This problem was overcome in 1909 when the Danish chemist Soren Sorenson proposed the pH scale (p stands for 'potenz' the German for strength). He defined pH as:

$$pH = -\log[H^+]$$ where $[H^+]$ is the concentration of H^+ in $mol \, dm^{-3}$

The negative sign in the equation results in pH decreasing as the aqueous hydrogen ion concentration increases.

Key Terms

A strong acid is one that fully dissociates in aqueous solution.

A weak acid is one that partially dissociates in aqueous solution.

46

Knowledge check

Explain why nitric acid is classified as a strong acid.

47

Knowledge check

Differentiate clearly between a strong acid and a concentrated acid.

YOU SHOULD KNOW ›››

››› the pH scale (prior knowledge)

››› the relationship between pH and H^+(aq)

››› how to calculate pH from $[H^+]$ and vice versa

Knowledge check

Calculate:

(a) The pH of a solution with an aqueous hydrogen ion concentration of $0.01\,mol\,dm^{-3}$.

(b) The aqueous hydrogen ion concentration of a solution of pH 2.5.

▼ Study point

The simplest way to measure pH is by using universal indicator. More accurate measurements can be made by using pH meters.

▼ Study point

The higher the H^+ ion concentration, the lower the pH and the stronger the acid.

Worked examples

What is the pH of:

(a) a sample of rain water with a $H^+(aq)$ concentration of $3.9 \times 10^{-6}\,mol\,dm^{-3}$?

$$pH = -\log(3.9 \times 10^{-6})$$
$$pH = 5.4$$

(b) A solution with a $H^+(aq)$ concentration of $1.0\,mol\,dm^{-3}$?

$$pH = -\log 1$$
$$pH = 0$$

Sometimes it is useful to determine aqueous hydrogen ion concentrations from pH values, e.g. a sample of acid rain has a pH of 2.2. What is the aqueous hydrogen ion concentration of this sample?

$$pH = -\log[H^+]$$

multiply both sides by -1 and rearrange

$$\log[H^+] = -pH$$
$$\log[H^+] = -2.2$$

take the antilogarithm of both sides (often 'shift log' on a calculator)

$$[H^+] = 10^{-2.2}$$
$$[H^+] = 6.3 \times 10^{-3}\,mol\,dm^{-3}$$

Using the pH scale the acidity of any solution can be expressed as a simple more manageable number, ranging from 0 to 14. This is much more convenient for the general public when dealing with concepts of acidity.

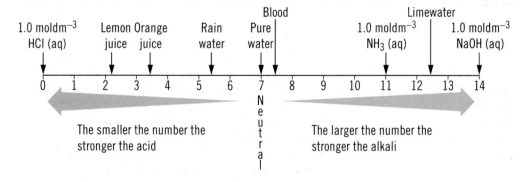

YOU SHOULD KNOW ⟩⟩⟩

⟩⟩⟩ the reactions of dilute acids with bases and carbonates (prior knowledge)

Key Term

A **salt** is the compound that forms when a metal ion replaces the hydrogen ion in an acid.

Neutralisation

The reactions of acids with bases and carbonates are not specifically mentioned in the specification. However, since all learners are expected to demonstrate knowledge and understanding of standard content covered at GCSE level, this section gives a recap on the minimum knowledge that is required about these reactions.

Bases (including alkalis) and carbonates react with acids in neutralisation reactions. The base (or carbonate) accepts the H^+ ions donated by the acid and the H^+ ions of the acid are replaced by metal ions (or NH_4^+ ions) to form a **salt**. For example, aqueous hydrochloric acid and aqueous sodium hydroxide react according to the equation:

$$HCl(aq) + NaOH(aq) \longrightarrow NaCl(aq) + H_2O(l)$$

If we write an ionic equation, we get:

$$H^+(aq) + Cl^-(aq) + Na^+(aq) + OH^-(aq) \longrightarrow Na^+(aq) + Cl^-(aq) + H_2O(l)$$

Removing the spectator ions gives:

$$H^+(aq) + OH^-(aq) \longrightarrow H_2O(l)$$

If an insoluble metal oxide, e.g. MgO, reacts with an acid, e.g. H_2SO_4, we get:

$$MgO(s) + H_2SO_4(aq) \longrightarrow MgSO_4(aq) + H_2O(l)$$

The oxide ion, O^{2-}, has accepted the H^+ ion to form water and the salt magnesium sulfate is formed.

If a carbonate, e.g. $PbCO_3$, reacts with an acid, e.g. HNO_3, we get:

$$PbCO_3(s) + 2HNO_3(aq) \longrightarrow Pb(NO_3)_2(aq) + H_2O(l) + CO_2(g)$$

This time the carbonate ion has accepted the H^+ ion to form water and carbon dioxide and the salt lead nitrate is formed.

Neutralisation always produces a salt. Many salts are very useful and neutralising an acid is a convenient way to form salts.

The same method can be used to form a salt from an insoluble base or carbonate but a different method has to be used to form a salt from an alkali since alkalis are soluble in water.

The method for preparing a salt from an acid and an alkali (or a soluble carbonate) is known as titration. (Details are shown on pages 72–73.)

A volume of alkali is measured into a flask and a few drops of indicator are added. Acid is added from a burette until the indicator changes colour. When the volume of acid needed to neutralise the alkali has been calculated, the procedure is repeated without the indicator so the correct amount of acid is added to the flask. The solution from the flask is heated to evaporate some of the water. Then it is left to cool and form crystals of pure salt.

Acid–base titrations

An acid–base titration is a type of volumetric analysis where the volume of one solution, say, an acid, that reacts exactly with a known volume of another solution, say, a base, is measured. The precise point of neutralisation is measured using an indicator. Titrations are not just used for preparing salts, they are often used for calculating the exact concentrations of acid or base solutions. To do this one of the solutions must be a **standard solution** or it must have been standardised. In the analysis you use the standard solution to find out information about the substance dissolved in the other solution.

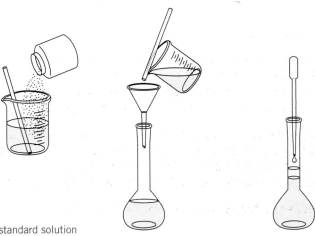
▲ Preparing a standard solution

To calculate the mass of reagent required to make a standard solution use the expressions $n = cv$ and $m = n \times M$. For example, what mass of anhydrous sodium carbonate must be dissolved in 250 cm^3 of solution to obtain a concentration of 0.0500 mol dm^{-3}?

$$n(Na_2CO_3) = 0.250 \times 0.0500$$
$$= 1.25 \times 10^{-2} \text{ mol dm}^{-3}$$
$$\text{mass } Na_2CO_3 = 1.25 \times 10^{-2} \times 106$$
$$= 1.325 \text{ g}.$$

Standard solution

A standard solution is prepared using a primary standard. A primary standard is typically a reagent which can be weighed easily, and which is so pure that its weight is truly representative of the number of moles of substance contained. Features of a primary standard include:

1. High **purity**
2. Stability (low **reactivity**)
3. Low **hygroscopicity** (to minimise weight changes due to humidity)
4. High **molar** mass (to minimise weighing errors).

Sodium hydroxide cannot be used as a primary standard because it reacts with atmospheric carbon dioxide.

Two examples of primary standards are:

- **Potassium hydrogen phthalate** (usually called KHP) for standardisation of aqueous base solutions.
- **Sodium carbonate** for standardisation of aqueous acids.

A standard solution is prepared from a solid as follows:

- Calculate the mass of the solid required and accurately weigh this amount into a weighing bottle.
- Transfer all of the solid into a beaker. Wash out the weighing bottle so that all the weighings run into the beaker. Add water and stir until all the solid dissolves.
- Pour all the solution carefully through a funnel into a volumetric (graduated) flask, washing all the solution out of the beaker and the glass rod. Add water until just below the graduation mark.
- Add water drop by drop until the graduation mark is reached and mix the solution thoroughly.

Performing a titration

All titrations involve the same major equipment / chemicals, namely:

- A burette containing one solution (e.g. an acid).
- A conical flask containing the other solution (e.g. a base).
- A pipette to accurately transfer the solution to the conical flask.
- An indicator to show when the reaction is completed.

(Two common indicators are phenolphthalein which is colourless in acid solution and purple in alkaline solution and methyl orange which is red in acid solution and yellow in alkaline solution.)

All titrations follow the same overall method:

- Pour one solution, say, an acid, into a **burette**, using a funnel, making sure that the jet is filled. Remove the funnel and read the burette.
- Use a **pipette** to add a measured volume of the other solution, say, a base, into a **conical flask**.
- Add a few drops of indicator to the solution in the flask.
- Run the acid from the burette to the solution in the conical flask, swirling the flask.
- Stop when the indicator just changes colour (this is the end-point of the titration).
- Read the burette again and subtract to find the volume of acid used (this is known as the titre).

Two important terms in titration are 'end point' and 'equivalence point'. The end point of a titration is when the indicator changes colour, i.e. it is a property of the indicator, it does not mean that the reaction is complete. The equivalence point of the titration occurs when the two solutions have reacted exactly, i.e. moles acid = moles base.

There are many different acid–base indicators and it is important to choose the correct one for a particular reaction. If the correct indicator is chosen, the end point of the titration will coincide exactly with its equivalence point. Also the indicator should show a distinct colour change.

▲ The burette is filled with acid using a funnel

▲ A graduated pipette is used to measure 25 cm³ of alkali into a conical flask

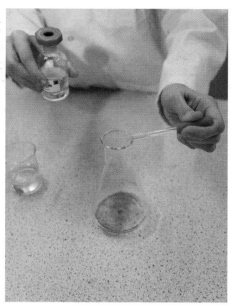

▲ An indicator e.g. phenolphthalein is added to the alkali which turns purple

▲ Acid is added to the alkali, swirling the flask continually

▲ The indicator has changed colour showing that the end-point has been reached

- Repeat the titration, making sure that the acid is added drop by drop near the end point, until you have at least two readings that are within 0.20 cm³ of each other and calculate a mean titre.

Double titrations

Since different indicators change colour at different pH values, if a solution contains a mixture of two bases which are of different strengths, one titration can be performed, but in two stages, using two different indicators, one added at each stage, to calculate the concentrations of both bases.

For example, the concentrations of sodium hydroxide, NaOH, and sodium carbonate, Na_2CO_3, in a mixture can be determined by titrating with hydrochloric acid, HCl, using phenolphthalein and methyl orange as indicators.

PRACTICAL CHECK

Standardising an acid solution is a **specified practical task**.

In the preparation of a standard solution:

Use the tare button on the weighing balance so that the scale reads zero.

When adding solid to the weighing bottle, remove the bottle from the pan and then add the solid, checking the mass until the correct amount has been added. This prevents errors caused by spilling solid onto the pan of the balance.

In the titration:

When using a pipette, always use a pipette filler – never suck the liquid up by mouth.

Never blow the contents of a pipette into the conical flask.

The conical flask can be wet as this will not alter the number of moles of alkali/acid added from the pipette, but the flask must not contain any acid or alkali.

The burette should be rinsed out with a few cm³ of the appropriate acid/alkali.

Record the volume in the burette by looking at where the bottom of the meniscus is on the scale.

Always estimate the volume to the nearest 0.05 cm³.

PRACTICAL CHECK

You should be able to standardise an acid solution.

You should be able to perform a double titration.

You should be able to perform a back titration.

▼ **Study point**

Make sure that you can explain why each step is carried out in a titration.

50

Knowledge check

Give two reasons why sodium hydroxide is unsuitable as a primary standard to prepare a standard solution.

51

Knowledge check

During an acid–base titration, explain why:

(a) A measuring cylinder is not used to transfer the basic solution into a conical flask.

(b) An indicator is added to the basic solution in the conical flask.

Phenolphthalein changes colour at around pH 9 (pink to colourless).

Methyl orange changes colour at around pH 4 (yellow to orange).

Sodium hydroxide is neutralised by the acid according to:

$$OH^-(aq) + H^+(aq) \longrightarrow H_2O(l)$$

When sodium carbonate is neutralised by the acid two reactions occur:

$$CO_3^{2-}(aq) + H^+(aq) \longrightarrow HCO_3^-(aq)$$

$$HCO_3^-(aq) + H^+(aq) \longrightarrow H_2O(l) + CO_2(g)$$

However, at pH 9 all the hydroxide ions have been neutralised and the carbonate ions have been converted to hydrogencarbonate ions. At pH 4 the hydrogencarbonate ions are converted to water and carbon dioxide.

i.e.
$$\left. \begin{array}{l} OH^-(aq) + H^+(aq) \longrightarrow H_2O(l) \\ CO_3^{2-}(aq) + H^+(aq) \longrightarrow HCO_3^-(aq) \end{array} \right\} \begin{array}{l} \text{both completed at} \\ \text{phenolphthalein stage} \end{array}$$

$$HCO_3^-(aq) + H^+(aq) \longrightarrow H_2O(l) + CO_2(g) \qquad \text{completed at methyl orange stage}$$

Therefore the first stage of the titration (change in colour of the phenolphthalein) relates to the concentration of the hydroxide and the carbonate.

The second stage of the titration (after addition of methyl orange) relates to the concentration of carbonate only. (Since $1\,mol\ HCO_3^- = 1\,mol\ CO_3^{2-}$).

Worked example

$25.0\,cm^3$ of a solution containing sodium hydroxide and sodium carbonate were titrated against dilute hydrochloric acid of concentration $0.100\,mol\,dm^{-3}$ using phenolphthalein as indicator. After $22.00\,cm^3$ of acid had been added the indicator was decolourised. Methyl orange was added and a further $8.25\,cm^3$ of acid were needed to turn the indicator orange.

Calculate the concentrations of sodium hydroxide and sodium carbonate in the solution.

$$\text{Moles HCl (in first stage)} = \frac{22.00}{1000} \times 0.100 = 2.20 \times 10^{-3}\,mol$$

Therefore $\qquad \text{moles } OH^- + CO_3^{2-} = 2.20 \times 10^{-3}\,mol$

$$\text{Moles HCl (in second stage)} = \frac{8.25}{1000} \times 0.100 = 8.25 \times 10^{-4}\,mol$$

Therefore $\qquad \text{moles } HCO_3^- = 8.25 \times 10^{-4}\,mol$

Since $\qquad 1\,mol\ HCO_3^- = 1\,mol\ CO_3^{2-}$

$\qquad \text{Moles } CO_3^{2-} = 8.25 \times 10^{-4}\,mol$

$\qquad \text{Moles } OH^- = 2.20 \times 10^{-3} - 8.25 \times 10^{-4} = 1.375 \times 10^{-3}\,mol$

$$\text{Concentration } CO_3^{2-} = \frac{8.25 \times 10^{-4}}{0.025} = 0.0330\,mol\,dm^{-3}$$

$$\text{Concentration } OH^- = \frac{1.375 \times 10^{-3}}{0.025} = 0.0550\,mol\,dm^{-3}$$

Back titrations

Sometimes it is not possible to use standard titration methods. For example, the reaction between determined substance and titrant can be too slow, there can be a problem with end point determination or the base is an insoluble salt. In such situations we often use a technique called back titration.

In back titration a known excess of one reagent **A** reacts with an unknown amount of reagent **B**. At the end of the reaction, the amount of reagent **A** that remains is found by titration. A simple calculation gives the amount of reagent **A** that has been used and the amount of reagent **B** that has reacted.

Worked example

A 1.00 g sample of limestone is allowed to react with 100 cm³ of dilute hydrochloric acid of concentration 0.200 mol dm⁻³ for neutralisation. The excess acid required 22.8 cm³ of aqueous sodium hydroxide solution of concentration 0.100 mol dm⁻³.

Calculate the percentage of calcium carbonate in the limestone sample.

Equation for titration is given by:

$$NaOH(aq) + HCl(aq) \longrightarrow NaCl(aq) + H_2O(l)$$

$$\text{Moles NaOH} = \frac{22.8}{1000} \times 0.100 = 2.28 \times 10^{-3}\,\text{mol}$$

From equation   1 mol NaOH = 1 mol HCl

$$\text{Moles HCl in excess} = 2.28 \times 10^{-3}\,\text{mol}$$

Equation for reaction between limestone and acid is given by:

$$CaCO_3(s) + 2HCl(aq) \longrightarrow CaCl_2(aq) + H_2O(l) + CO_2(g)$$

$$\text{Moles HCl added} = \frac{100}{1000} \times 0.200 = 0.0200\,\text{mol}$$

$$\text{Moles acid reacted} = 0.0200 - 2.28 \times 10^{-3} = 0.01772\,\text{mol}$$

From equation 2 mol HCl = 1 mol $CaCO_3$

$$\text{Moles } CaCO_3 = \frac{0.01772}{2} = 8.86 \times 10^{-3}\,\text{mol}$$

$$\text{Mass } CaCO_3 = 8.86 \times 10^{-3} \times 100 = 0.886\,\text{g}$$

$$\text{\%age } CaCO_3 = \frac{0.886}{1.00} = 88.6\%$$

PRACTICAL CHECK

Performing a back titration and a double titration are **specified practical tasks**.

Exact details of titrations may differ, but the procedures will remain the same.

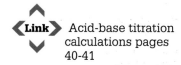 **Link** Acid-base titration calculations pages 40-41

1.1

1   Given that the formula for the phosphate ion is PO_4^{3-}, write the formula of calcium phosphate.   [1]

2   A metal M forms a sulfate with the formula MSO_4. Write the formula of its hydroxide.   [1]

3   State the oxidation number of chromium in CrO_2Cl_2.   [1]

4   Magnetite ore can be reduced by carbon monoxide in a blast furnace to produce iron as part of steel production.
Balance the equation for the reduction of magnetite.   [1]
$$Fe_3O_4 + CO \longrightarrow Fe + CO_2$$

5   When ethanol, C_2H_5OH, is burned in an excess of air, the only products are carbon dioxide and water.
Balance the equation for this reaction.   [1]
$$C_2H_5OH + O_2 \longrightarrow CO_2 + H_2O$$

6   When calcium is added to cold water, calcium hydroxide and hydrogen form.
Write the balanced chemical equation for this reaction.   [1]

7   Write a balanced equation for the reaction between PCl_5 and water to form H_3PO_4 and HCl only.   [1]

8   Write an ionic equation, including state symbols, for the reaction between hydrochloric acid and magnesium to form aqueous magnesium chloride and hydrogen.   [1]

9   Methane reacts with hot copper(I) oxide to produce copper, carbon dioxide and steam.
Write the balanced chemical equation for this reaction.   [1]

10 Magnesium nitride, Mg_3N_2, reacts with water to form magnesium hydroxide and ammonia.
Write the balanced chemical equation for this reaction.   [1]

11. One of the main uses of ammonia is in the production of nitric acid. In the first part of this process a mixture of ammonia and air is passed over a catalyst at 850 °C.
$$NH_3(g) + O_2(g) \longrightarrow NO(g) + H_2O(g)$$

    (a) Balance the equation above.   [1]

    (b) Complete the table below to give the oxidation numbers (states) of each element present.   [3]

Element	Initial oxidation number	Final oxidation number
Nitrogen		
Hydrogen		
Oxygen		

12 When an aqueous solution of calcium hydroxide is added to an aqueous solution of sodium carbonate a white precipitate of calcium carbonate is seen.

Write the ionic equation for this reaction. Include the relevant state symbols in the equation. [1]

1.2

1 **(a)** Name the particles contained in atomic nuclei and give their relative masses and charges by completing the table. [2]

Particle	Relative mass	Relative charge

(b) An element, X, has an atomic number of 9 and forms an ion X⁻.

State the number of protons and electrons in this ion. [1]

(c) The symbols $^{35}_{17}Cl$, $^{37}_{17}Cl$ and $^{39}_{19}K$, represent chlorine atoms and potassium atoms respectively.

Use these symbols to explain the meaning of the terms:

(i) Atomic number [1]

(ii) Isotope [1]

(d) Sketch a diagram to show the shape of a p orbital. [1]

(e) Show the electronic configuration of a potassium atom. [1]

2 **(a)** Atoms which have unstable nuclei are radioactive.

All living organic matter contains carbon-14, which emits β-particles and has a half-life of 5730 years.

(i) Explain what is meant by the term *half-life* of a radioactive isotope. [1]

(ii) Give the symbol of the element produced by the radioactive decay of ^{14}C, showing both its mass and atomic numbers. [1]

(iii) State what happens in the nucleus of an atom when a β-particle forms. [1]

(iv) Calculate how long it will take a sample of ^{14}C of mass 8g to decay to 1g. [1]

(b) Some information relating to three other radioisotopes is given in the table.

Isotope	Half-life	Radioactive emission
^{60}Co	5 years	γ
^{63}Ni	100 years	β
^{66}Cu	30 seconds	β

Use all the information to suggest which radioisotope would be the most suitable for use in a gauge to measure the thickness of aluminium foil. Explain your reasoning. [2]

3 **(a)** Magnesium is best known for burning with a characteristic brilliant white light; however, in industry it is the third most commonly used structural metal. The metal itself was first produced by Sir Humphry Davy in 1808 using electrolysis of a mixture of magnesia and mercury oxide.

Magnesium has three stable isotopes ^{24}Mg, ^{25}Mg and ^{26}Mg.

(i) State the number of protons present in an atom of ^{24}Mg. [1]

(ii) Deduce the number of neutrons present in an atom of ^{26}Mg. [1]

(b) Magnesium also has a radioactive isotope ^{28}Mg which has a half-life of 21 hours.

If you started with 2.0g, calculate the mass of ^{28}Mg remaining after 84 hours. [1]

(c) Name one useful radioactive isotope and describe how it is used. [2]

(d) Radon is a noble gas. Its most stable isotope ^{222}Rn has a half-life of 3.8 days, decays by α-emission and is responsible for the majority of the public exposure to ionising radiation.

(i) Give the symbol and mass number of the atom formed by the loss of one α-particle from an atom of ^{222}Rn. [1]

(ii) Explain why doctors are concerned that an over-exposure to radon may cause lung cancer. [1]

4 (a) Potassium-40, $^{40}_{19}$K, is a radioactive isotope that decays by β-emission and has a half-life of 1.25×10^9 years.

(i) Write an equation for the process by which a potassium-40 isotope emits a β-particle. [2]

(ii) Calculate how long it will take for the activity of the isotope to decay to $\frac{1}{8}$th of its original activity. [1]

(b) Four radioactive isotopes with the same mass are given in the table.

Isotope	Half-life	Radioactive emission
^{190}W	30 minutes	β
^{190}Re	3.1 minutes	β
^{190}Pt	6.5×10^{11} years	α
^{190}Bi	6.3 seconds	α

Describe why radioactivity is dangerous to living cells. Use all the data given to identify which of these isotopes would cause most damage to cells if consumed. [4]

5 (a) The electronic structures of five atoms are listed below. Arrange these atoms in order of increasing molar first ionisation energy. [2]

Atom	A	B	C	D	E
Electronic structure	$1s^2$	$1s^22s^2$	$1s^22s^22p^1$	$1s^22s^22p^3$	$1s^22s^22p^6$

(b) State, giving a reason for your choice, which one of the following gives the first four ionisation energies for silicon, Si. [2]

	Ionisation energy / kJ mol^{-1}			
	1st	2nd	3rd	4th
A	496	4563	6913	9544
B	578	1817	2745	11578
C	738	1451	7733	10541
D	789	1577	3232	4356

6 (a) Hydrogen has a first ionisation energy of 1312 kJ mol^{-1}.

Explain why helium has a higher first ionisation energy than hydrogen. [2]

(b) Sodium and potassium are both in group 1 of the periodic table.

Explain why sodium has a higher first ionisation energy than potassium. [2]

(c) The table below gives the first three ionisation energies for boron and potassium:

Element	Ionisation energy / kJ mol^{-1}		
	1st	2nd	3rd
B	800	2420	3660
K	419	3051	4412

(i) Suggest why compounds containing B^{3+} ions are unlikely to exist. [1]

(ii) Write an equation to represent the second ionisation energy of potassium. [1]

(iii) State how the first three ionisation energies of calcium would differ from those of potassium. [2]

7 The equation below shows the removal of 2 electrons from an element X

$$X(g) \longrightarrow X^{2+}(g) + 2e^-$$

where X is calcium, magnesium or sodium.

The table below shows the energy required to remove 2 electrons from each of these elements.

X	Energy required to remove 2 electrons / kJ mol⁻¹
Ca	1735
Mg	2189
Na	5059

Explain, in terms of the electronic structures of the atoms concerned, why

(a) the energy required to remove 2 electrons from calcium is less than for magnesium [2]

(b) the energy required to remove 2 electrons from sodium is greater than for magnesium. [2]

8 **(a)** The diagram below shows the emission spectrum of the hydrogen atom in the visible region.

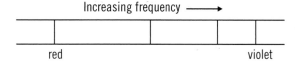

(i) Explain why hydrogen emits only certain definite frequencies of visible light. [2]

(ii) The horizontal lines in the diagram below show the electron energy levels of a hydrogen atom.

On a copy of the diagram label these horizontal lines and draw the transitions corresponding to the four spectral lines in (a) above, clearly indicating which transition represents the red spectral line. [3]

(iii) On a copy of the diagram draw and label the transition corresponding to the ionisation of the atom. [1]

(b) The first line in the visible atomic emission spectrum for hydrogen has a wavelength of 656 nm, while that for helium has a wavelength of 707 nm.

State, giving a reason, which line has:

(i) The higher frequency. [1]

(ii) The higher energy. [1]

(c) (i) A line in the ultra-violet region of the electromagnetic spectrum (part of the Lyman series) of hydrogen has a wavelength of 9.12×10^{-8} m.

Use the equation below to calculate the frequency, f, of this line and hence its energy, E, in joules. Show your working. [3]

$$c = f \times \lambda$$

(Assume that $c = 2.99 \times 10^8$ m s^{-1}, $h = 6.63 \times 10^{-34}$ J s)

(ii) There are 6.02×10^{23} hydrogen atoms in 1 mol. Use this information and your answer to part (i) to calculate the energy in kJ mol^{-1}. [2]

9 This question is about atomic structure.

(a) Give the full electronic configuration of a nitrogen atom and use this to describe the way in which electrons are arranged in atoms. [4]

(b) Describe the main features of the atomic emission spectrum of hydrogen in the visible region. Explain how these features arise and how their interpretation provides evidence for energy levels in the atom. [6]

1.3

1 Calculate the mass of silver that contains the same number of atoms as there are molecules in 11.0 g of carbon dioxide, CO_2. [1]

2 (a) Calculate the molar mass, in g mol^{-1}, of calcium sulfate dihydrate, $CaSO_4.2H_2O$. [1]

(b) Calculate the percentage of water, by mass, in calcium sulfate dihydrate. [1]

3 Potassium metal was discovered in 1807 by the British chemist Sir Humphrey Davy. Its name derives from the word 'potash' since potassium was isolated by the electrolysis of molten caustic potash, KOH.

(a) The mass spectrum of a naturally occurring sample of potassium gave the following results:

Isotope	% abundance
^{39}K	93.26
^{40}K	0.012
^{41}K	6.730

These results can be used to determine the relative atomic mass of the potassium sample.

(i) Complete the following definition of *relative atomic mass*: [1]

The relative atomic mass of an element is the average mass of one atom of the element relative to …

(ii) Calculate the relative atomic mass of the sample, giving your answer to four significant figures. [2]

(b) This mass spectrum was produced by potassium ions in a mass spectrometer.

State why a vacuum is needed inside the apparatus. [1]

4 (a) A sample of water containing $^1H_2$, ^{16}O and $^2H_2{}^{16}O$ was analysed in a mass spectrometer. The trace showed a number of peaks. Suggest which ions are responsible for the peaks at mass number 2, 18 and 20. [2]

(b) On earth, iodine occurs as only one stable isotope, ^{127}I.

(i) Using your understanding of the mass spectrum of chlorine, Cl_2, sketch and label the mass spectrum of a sample of iodine. [2]

(ii) A sample of iodine extracted from a meteorite was found to have a relative atomic mass of 128.7. State what this tells you about the composition of the iodine within the meteorite. [1]

5 When a sample of $^{210}PoCl_4$ was passed through a mass spectrometer, the spectrum showed five molecular ion peaks corresponding to masses 350, 352, 354, 356 and 358.

(a) State the two mass numbers of the isotopes of chlorine found in naturally occurring chlorine. [1]

(b) Each of the five molecular ions giving peaks in the mass spectrum of $^{210}PoCl_4$ contains one ^{210}Po atom and four Cl atoms. Using your answer to (a), state how many atoms of each of the two Cl isotopes are present in each molecular ion. [2]

(c) Explain how the positive ions produced in a mass spectrometer are formed and separated. [3]

6 A chlorate of sodium which is used to bleach pulp has a percentage composition, by mass, of Na 21.6%; Cl 33.3%; O 45.1%.

(a) Calculate the **empirical** formula of this compound. [2]

(b) What other information would you need to know to be able to deduce the **molecular** formula of this compound? [1]

7 Sodium nitrate is widely used in the production of fertilisers.

(a) Sodium nitrate can be formed by the reaction between sodium carbonate and nitric acid as shown by the equation

$$Na_2CO_3(aq) + 2HNO_3(aq) \longrightarrow 2NaNO_3(aq) + H_2O(l) + CO_2(g)$$

In an experiment, $25.0\,cm^3$ of a solution of sodium carbonate of concentration $0.0450\,mol\,dm^{-3}$ required $23.6\,cm^3$ of the acid for complete neutralisation.

Calculate the concentration of the acid, in $mol\,dm^{-3}$. [3]

(b) Sodium nitrate decomposes on heating as shown by the equation

$$2NaNO_3(s) \longrightarrow 2NaNO_2(s) + O_2(g)$$

When a sample was decomposed completely, the oxygen occupied a volume of $700\,cm^3$ at a pressure of $101\,kPa$ and a temperature of $28\,°C$.

Calculate the amount, in moles, of oxygen produced. [3]

(The gas constant $R = 8.31\,J\,K^{-1}\,mol^{-1}$.)

8 Hannah is asked to measure the rate of reaction of calcium carbonate with dilute hydrochloric acid. She is given $1.50\,g$ of the carbonate and $20.0\,cm^3$ of hydrochloric acid of concentration $1.20\,mol\,dm^{-3}$.

$$CaCO_3(s) + 2HCl(aq) \longrightarrow CaCl_2(aq) + CO_2(g) + H_2O(l)$$

(a) Calculate the number of moles of hydrochloric acid used in this reaction. [1]

(b) Calculate the **minimum** mass of calcium carbonate needed to react **completely** with this amount of acid. [2]

(c) Calculate the volume of carbon dioxide that would be produced at $25\,°C$. [2]

(1 mole of carbon dioxide occupies $24.0\,dm^3$ at $25\,°C$.)

(d) Calculate the volume of carbon dioxide that would be produced at $50\,°C$. [2]

(Assume that the pressure remains unchanged.)

9 Sodium carbonate can be manufactured in a two-stage process as shown by the following equations:

$NaCl + NH_3 + CO_2 + H_2O \longrightarrow NaHCO_3 + NH_4Cl$ (1)

$2NaHCO_3 \longrightarrow Na_2CO_3 + H_2O + CO_2$ (2)

(a) Calculate the atom economy for the production of sodium hydrogencarbonate in equation (1). [2]

(b) Use both equations to calculate the maximum mass of sodium carbonate which could be obtained from 900g of sodium chloride. [3]

10 Epsom salts is a form of hydrated magnesium sulfate. It can be represented by the formula $MgSO_4.xH_2O$.

When 7.38g of this hydrate was heated, 3.60g of the anhydrous salt, $MgSO_4$, remained.

Calculate the value of x in $MgSO_4.xH_2O$ [3]

11 This question involves two different methods to determine the percentage of sodium carbonate in a mixture.

(a) Elinor is given the mixture and she carries out a titration to determine the percentage of sodium carbonate in the mixture.

She dissolves 2.05g of the sample in distilled water and accurately makes up the solution to $250\,cm^3$ in a volumetric flask.

She pipettes $25.0\,cm^3$ of the solution into a conical flask, adds an appropriate indicator and titrates this solution with hydrochloric acid of concentration $0.100\,mol\,dm^{-3}$. She repeats the procedure three times and finds that she needs $23.15\,cm^3$ of acid to react completely with the sodium carbonate.

The equation for the reaction is given below

$Na_2CO_3 + 2HCl \longrightarrow 2NaCl + H_2O + CO_2$

   (i) Calculate the number of moles of HCl used in the titration and hence deduce the number of moles of Na_2CO_3 in $25.0\,cm^3$ of the solution. [2]

   (ii) Calculate the percentage of Na_2CO_3 in the mixture. [3]

(b) In a separate experiment Edmund was asked to analyse the mixture by means of a precipitation reaction.

Edmund measured exactly 2.1g of the mixture and dissolved it in an excess of distilled water. To this solution he added an excess of barium nitrate solution. A precipitate of barium carbonate was formed according to the equation:

$Na_2CO_3(aq) + Ba(NO_3)_2(aq) \longrightarrow BaCO_3(s) + 2NaNO_3(aq)$

The precipitate was filtered and the mass of barium carbonate formed was found to be 2.3g.

   (i) Explain why the volume of distilled water used to dissolve the original mixture was not measured accurately. [1]

   (ii) Use the mass of barium carbonate formed to calculate the mass of Na_2CO_3 in the original mixture and hence the percentage of Na_2CO_3 in the mixture. [3]

(c) (i) Explain which of the two methods in parts (a) and (b) is likely to give the more accurate result. [1]

   (ii) Suggest possible improvements to the method giving the less accurate result. [2]

12 Berian was asked to find the identity of a group 1 metal hydroxide by titration. He dissolved 1.14 g of the hydroxide in water, and made up the solution to 250 cm^3. He accurately transferred 25.0 cm^3 of this solution into a conical flask. The solution required 23.80 cm^3 of hydrochloric acid for complete neutralisation. The concentration of the acid solution was 0.730 g HCl in 100 cm^3 of water.

The equation for the reaction between the metal hydroxide and hydrochloric acid is given below. M represents the symbol of the metal.

$$MOH + HCl \longrightarrow MCl + H_2O$$

(a) Calculate the number of moles of HCl used in the titration. [3]

(b) Deduce the number of moles of MOH in 25.0 cm^3 of the solution and hence calculate the total number of moles of MOH in the original solution. [2]

(c) Calculate the relative molecular mass of MOH and hence deduce the metal in the hydroxide. [2]

13 When 1 kg of sulfur dioxide is reacted with excess oxygen, 1.225 kg of sulfur trioxide is formed:

$$2SO_2(g) + O_2(g) \longrightarrow 2SO_3(g)$$

Calculate the percentage yield. [3]

14 Limestone contains calcium carbonate. A 0.497 g sample of ground limestone was placed in a flask and exactly 25.0 cm^3 of hydrochloric acid of concentration 0.515 mol dm^{-3} was added. The mixture was stirred until no more bubbles of carbon dioxide were formed. The unreacted acid in the flask was titrated against 0.188 mol dm^{-3} sodium hydroxide and required 24.80 cm^3 for neutralisation.

(a) Calculate the number of moles of hydrochloric acid used up in the reaction with limestone. [3]

(b) Calculate the percentage of calcium carbonate in the limestone sample.

$$CaCO_3 + 2HCl \longrightarrow CaCl_2 + H_2O + CO_2$$ [3]

1.4

1 Draw a dot and cross diagram to show how the ionic compound calcium chloride is formed from calcium and chlorine atoms. Show the charges on the ions formed.

Only outer electrons should be shown. [2]

2 Aluminium chloride forms a dimer, Al_2Cl_6, that contains both covalent bonds and coordinate bonds. Describe what is meant by the terms covalent bond and coordinate bond. [2]

3 The electronegativity values of some elements are given below:

Atom	H	N	O	Al	Cl	F
Electronegativity value	2.1	3.0	3.5	1.6	3.0	4.0

(a) Use this data to identify any dipoles present in the following bonds. Mark their polarity clearly.

N—H   O—Cl [1]

(b) Use the data to explain why aluminium chloride is considered to be a covalent compound while aluminium oxide is anionic compound. [1]

4 Chlorine fluoride has the following formula. Cl F

Cl—F

(a) Indicate the polarity in the bond shown by use of the symbols δ+ and δ–, giving a reason for your answer. [1]

(b) Draw a dot and cross diagram to illustrate the bonding between the two atoms in chlorine fluoride. Include **all** outer shell electrons. [1]

5 When the temperature is increased, both solid iodine and diamond change directly into their gaseous state – they sublime.

    (a) In each case, name the force or bond that is being overcome when the solid changes into a gas. [2]

        Iodine ....

        Diamond ....

    (b) State, with a reason, which solid would have the higher sublimation temperature. [1]

6 Put the following in order of increasing strength.

    covalent bonds   hydrogen bonds   van der Waals forces

    weakest ⎯⎯⎯⎯⎯⎯⎯⟶ strongest [1]

7 **(a)** Use the VSEPR theory to deduce the shapes of BF_3 and NH_3. [2]

    (b) Boron fluoride reacts with ammonia, NH_3, to make the compound shown in the following equation:

$$BF_3 + NH_3 \longrightarrow F_3B - NH_3$$

        (i) Name the type of bond formed between N and B. [1].

        (ii) Suggest a value for the F⎯B⎯F bond angle in this molecule.

        Explain your answer to part (ii). [2]

8 The bond angles in molecules of methane and water are shown in the diagrams below.

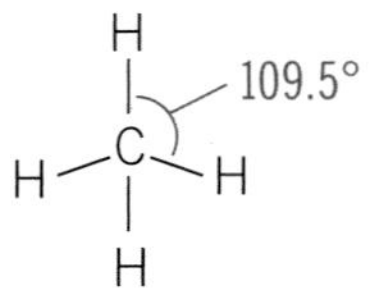

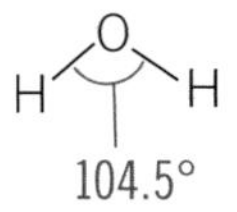

    Using the valence shell electron pair repulsion (VSEPR) theory:

    (i) State why methane has the shape shown. [2]

    (ii) Explain why the H⎯O⎯H bond angle in water is less than the H⎯C⎯H bond angle in methane. [2]

9 Aluminium, boron and nitrogen all form chlorides containing three chlorine atoms, XCl_3. Molecules of boron chloride, BCl_3, and molecules of nitrogen chloride, NCl_3, have different shapes.

    Use VSEPR (valence shell electron pair repulsion) theory to:

    (i) State, and [3]

    (ii) Explain the shapes of these molecules. [3]

10 Select all the molecules from the list below that have bond angles of less than 109°.

    NH_4^+ BF_3 NH_3 CH_4 SF_6 H_2O [2]

11 Sulfur difluoride dioxide (sulfuryl fluoride), SO_2F_2, is used as a gaseous insecticide to control termite infestations in wooden houses.

   (a) It can be produced by reacting together sulfur dioxide and fluorine.

$$SO_2 + F_2 \longrightarrow SO_2F_2$$

     Use the oxidation numbers of sulfur to show that sulfur has been oxidised in this reaction. In your answer you should state how changes in oxidation number are related to oxidation. [2]

   (b) Sulfuryl fluoride is a tetrahedral molecule where the sulfur atom has no lone pairs of electrons.

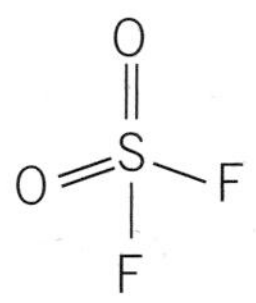

$$O \!=\!\! \underset{\underset{F}{|}}{\overset{\overset{O}{\|}}{S}} \!\!-\! F$$

     Use the valence shell electron pair repulsion theory (VSEPR) to explain why sulfuryl fluoride has this shape. [1]

12 The boiling temperatures of some hydrogen halides are shown in the table below:

Hydrogen halides	Boiling temperature / K
HF	293
HCl	188
HBr	206
HI	238

Explain, in terms of the nature of the intermolecular bonding present, why the boiling temperature of hydrogen iodide is higher than those of hydrogen chloride and hydrogen bromide but is lower than that of hydrogen fluoride. [4]

1.5

1 Lithium bromide has the same crystal structure as sodium chloride.

Label the ions present in the diagram of lithium bromide shown below.

(You may assume that the anion is larger than the cation.) [1]

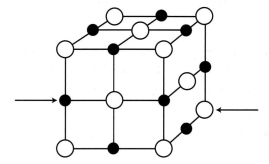

2 **(a)** Both sodium chloride and caesium chloride have giant ionic structures.

     (i) Draw a labelled diagram to show the arrangement of ions in a crystal of caesium chloride. [2]

     (ii) Give a reason why sodium chloride has a different structure from caesium chloride. [1]

(b) Diamond and graphite are two different crystalline forms of carbon.

    (i) Describe the structure and bonding in both diamond **and** graphite. [4]

    (ii) State **one** physical property which is common to both diamond and graphite and **one** which is not. Relate **both** properties to the structures and bonding you have described. [2]

3 Both diamond and iodine contain covalent bonding.

Explain why solid iodine can be converted into a vapour at a much lower temperature than diamond.

You should refer to the type(s) of bonding involved and how this bonding affects melting temperatures. Include a diagram if you wish.

4 An element has a melting temperature of 28°C and conducts electricity when it is a solid or a liquid.

State the feature present in the structure that allows it to conduct electricity.

5 Although diamond is a non-metal, it can show similar physical properties to metals.

  (a) Describe the structure and bonding in diamond and in metals. [3]

  (b) State **one** physical property which is common to both diamond and metals and one which is not. Relate **both** properties to the structures and bonding you have described. [3]

6 Haematite is an ore of iron that contains Fe_2O_3. Iron is extracted from this ore in a blast furnace.

  (a) Balance the equation for the extraction of iron from Fe_2O_3. [1]

    .... Fe_2O_3 + ... CO = ----- Fe + ... CO_2

  (b) Use oxidation states to show that the reaction in (a) is a redox reaction. [2]

  (c) A different oxide of iron is iron(II) oxide, FeO. The ions in this compound adopt an arrangement similar to that of sodium chloride.

    (i) Give the crystal co-ordination numbers for the ions in FeO. [1]

    (ii) Draw the arrangement of oxide ions around each iron ion. [1]

  (d) Carbon monoxide contains two covalent bonds and one coordinate bond. Explain what is meant by the terms covalent bond and coordinate bond, indicating the difference between them. [2]

  (e) Iron is a typical metal. Describe the bonding present in iron. Explain how it can conduct electricity and why it has a high melting temperature. [4]

1.6

1 Barium chloride is a highly toxic compound that is frequently used in the laboratory. Aqueous barium chloride can be used to test for sulfate ions in solution.

  (a) **(i)** Write an **ionic** equation for the reaction that occurs when aqueous barium chloride is added to a solution containing sulfate ions. [1]

    (ii) Give the observation expected for a positive result in this chemical test. [1]

  (b) A solution of barium chloride can be identified using separate tests for barium ions and chloride ions.

    (i) A flame test can be used to prove that the solution contains barium ions. State the flame colour that would be seen. [1]

    (ii) Give a chemical test to show that the solution contains chloride ions. Your answer should include the reagent(s) and expected observation(s). [2]

    Reagent(s)
    Observation(s)

2 Explain the fact that the melting temperature of sodium is much lower than the melting temperature of magnesium.

You should include reference to the type(s) of bonding involved and how this bonding affects melting temperatures. You may include a diagram if you consider it helpful.   [3]

3 The tables show some of the elements in the periodic table together with the melting temperature of the most important solid form of the element.

	I	II	III	IV	V	VI	VII	0
Element				C (diamond)				Ne
Melting temperature / K				3900				25
Element	Na	Mg	Al	Si	P	S	Cl	Ar
Melting temperature / K	371	922	933	1683	317	392	172	84

   (i) State which of the elements shown will **not** be solids at room temperature.   [1]

   (ii) Explain, in terms of bonding and structure, why silicon, Si, and carbon, C, have the highest melting temperatures.   [2]

   (iii) Explain why neon, Ne, and argon, Ar, have the lowest melting temperatures.   [2]

   (iv) Explain why the melting temperature increases for the sequence Na, Mg, Al, even though all three involve metallic bonding.   [2]

   (v) State why the melting temperature for Si is lower than that for C even though they involve the same bonding and structure.   [1]

   (vi) State why the melting temperature of argon, Ar, is higher than that for neon, Ne.   [1]

4 The elements in Group 7 in the periodic table can be described as p-block. State why these are described as p-block elements.   [1]

5 All halogens are oxidising agents.

   (i) Why are the halogens oxidising agents?   [1]

   (ii) State, giving a reason, which halogen is the strongest oxidising agent.   [1]

   (iii) $NaClO_3$ was used as a weedkiller. Give the oxidation state of chlorine in $NaClO_3$.   [1]

6 A student was given a solution of calcium bromide and asked to carry out the reactions shown in the diagram below.

Flame test ⟵ calcium bromide solution ⟶ aqueous chlorine

aqueous silver nitrate

   (i) State the colour given in the flame test.   [1]

   (ii) State what was seen when aqueous silver nitrate was added.   [1]

   (iii) Give the **ionic** equation for the reaction occurring in (ii).   [1]

   (iv) State what was seen when aqueous chlorine was added to the solution of calcium bromide.   [1]

   (v) Explain why chlorine reacted as described in (iv).   [2]

Your answer should include:

- The type of bonding and the species present in calcium bromide.

- The type of reaction occurring.

- Why chlorine is able to react in this way.

- An appropriate equation.

7 Name the white precipitate formed when an aqueous solution of barium nitrate, $Ba(NO_3)_2$ is mixed with an aqueous solution of potassium carbonate and write an ionic equation for its formation. [2]

8 Basic magnesium carbonate has the formula $MgCO_3.Mg(OH)_2.3H_2O$. On strong heating it decomposes giving thee products, two of which are carbon dioxide and water. Give the equation for the thermal decomposition of basic magnesium carbonate. [2]

9 You are given four unlabelled bottles but know that these contain the following four solutions:

Potassium carbonate

Sodium hydroxide

Barium chloride

Magnesium chloride

    (i) Predict what will happen when each of the four solutions is added to the others, and present this information below. [4]

    (ii) Name the white precipitate formed when magnesium chloride is mixed with potassium carbonate, and write an ionic equation for its formation. [2]

        Name of precipitate
        Ionic equation

    (iii) State all the results if the three tests below are separately given to identify the four solutions. [4]

        litmus test
        flame test
        addition of sodium sulfate solution.

1.7

1 Ammonia, a very important industrial product, is produced by the Haber process. It can be converted to the common fertiliser, ammonium sulfate, by reacting it with sulfuric acid.

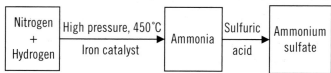

(a) **(i)** Write an equation for the acid–base reaction of ammonia with sulfuric acid. [1]

    (ii) Explain why ammonia behaves as a base in this reaction. [1]

(b) The Haber process can be represented by the following equation

$$N_2(g) + 3H_2(g) \rightleftharpoons 2NH_3(g) \qquad \Delta H = -92 \, kJ \, mol^{-1}$$

For the equilibrium reaction explain why:

    (i) A high pressure is used. [2]

    (ii) Ammonia is removed from the equilibrium mixture as it forms. [2]

(c) How would the equilibrium yield be affected if the reaction were run without using the catalyst? [1]

(d) Predict and explain the effect of increasing the temperature of the surroundings on the value of the equilibrium constant, K_c, for the reaction. [2]

2 **(a)** In sea water there are equilibria between carbon dioxide, hydrogencarbonate (HCO_3^-) ions and carbonate (CO_3^{2-}) ions.

$$CO_2(aq) + H_2O(l) \rightleftharpoons H^+(aq) + HCO_3^-(aq)$$

$$H^+(aq) + CO_3^{2-}(aq) \rightleftharpoons HCO_3^-(aq)$$

    (i) Use Le Chatelier's Principle to predict the effect on the first equilibrium and any change in pH when more carbon dioxide is dissolved. [2]

    (ii) State what would be the effect on the concentration of carbonate (CO_3^{2-}) ions of increasing the concentration of hydrogen (H^+) ions in the second equilibrium. [1]

(b) In fizzy drinks, carbon dioxide is dissolved in water under pressure and when the pressure is released the 'fizz' appears.

In a bottle of fizzy drink, the following chemical equilibrium exists:

$$CO_2(g) \rightleftharpoons CO_2(aq)$$

    (i) Chemical equilibria are often described as dynamic equilibria.

    Explain the term *dynamic equilibrium*. [1]

    (ii) When the top is removed from a bottle of fizzy drink it goes 'flat' because much of the dissolved carbon dioxide comes out of solution. Explain why this happens in terms of chemical equilibria. [2]

3 **(a)** Recently hydrogen has been receiving interest as a 'source of energy'. It can be prepared by the steam reforming of methane.

$$CH_4(g) + H_2O(g) \rightleftharpoons CO(g) + 3H_2(g) \qquad \Delta H = 206\,kJ\,mol^{-1}$$

    (i) Write an expression for the equilibrium constant, K_c, giving the units, if any. [2]

    (ii) State Le Chatelier's Principle. [1]

    (iii) State, giving your reasons, how the equilibrium yield of hydrogen is affected, if at all, by:

        I increasing the temperature at constant pressure [2]

        II increasing the pressure at constant temperature. [2]

(b) If carbon monoxide and hydrogen are passed over a catalyst at high temperatures and pressures, methanol can be produced.

$$CO(g) + 2H_2(g) \rightleftharpoons CH_3OH(g) \qquad \Delta H = -91\,kJ\,mol^{-1}$$

    (i) State how the equilibrium yield of methanol is affected by

        I an increase in temperature [1]

        II an increase in pressure. [1]

    (ii) Explain your answer to part (i). [2]

4 Complete the table which refers to the effect a change in conditions has on the position of equilibrium shown below. [3]

$$2SO_2(g) + O_2(g) \rightleftharpoons 2SO_3(g) \qquad \Delta H = -196\,kJ\,mol^{-1}$$

Change	Effect, if any, on position of equilibrium	Effect, if any, on value of K_c
Addition of reactant at constant temperature		
Drop in temperature	Shift to the right	

5   (a) Polluting gases such as sulfur dioxide, SO_2, produced from power stations, can cause the acidification of lakes far from the source of the pollution. At a lake-water pH of 6.0, water snails start to die and when the pH reaches 5.5, fish also begin to die.

    (i)   State how you would explain to the general public how the pH scale is used to describe levels of acidity. [2]

    (ii) Calculate the hydrogen ion concentration in the lake-water when fish start to die. [2]

  (b) An equation for the reaction of sulfur dioxide with water is shown below:

$$SO_2(aq) + H_2O(l) \rightleftharpoons H^+(aq) + HSO_3^-(aq)$$

    (i)   Use the equation to explain why sulfur dioxide is described as an acidic oxide. [1]

    (ii) Use Le Chatelier's principle to explain how the concentration of hydrogen ions, $H^+(aq)$, would change if more sulfur dioxide were dissolved in a solution that had reached dynamic equilibrium. [2]

  (c) Acids can be considered to be strong or weak and concentrated or dilute.

    For an aqueous solution of an acid, explain the difference between the meaning of the terms *weak acid* and *dilute acid*. [2]

6   Berian was asked to find the identity of a group 1 metal hydroxide by titration.

He was told to use the following method:

- Fill a burette with hydrochloric acid solution.
- Accurately weigh about 1.14 g of the metal hydroxide.
- Dissolve all the metal hydroxide in water, transfer the solution to a volumetric flask then add more water to make the final volume 250 cm³ of solution.
- Accurately transfer 25.0 cm³ of this solution into a conical flask.
- Add 2–3 drops of a suitable indicator to this solution.
- Carry out a rough titration of this solution with the hydrochloric acid.
- Accurately repeat the titration several times and calculate a mean titre.

  (a) Give a reason why Berian does not simply add 1.14 g of metal hydroxide to 250.0 cm³ of water. [1]

  (b) Name a suitable piece of apparatus for transferring 25.0 cm³ of the metal hydroxide solution to a conical flask. [1]

  (c) State why he adds an indicator to this solution. [1]

  (d) Suggest why Berian was told to carry out a rough titration first. [1]

  (e) Explain why he carried out several titrations and worked out a mean value. [1]

7   Frances measures a mass of hydrated copper(II) sulfate, $CuSO_4.5H_2O$, and uses this to make exactly 250.0 cm³ of copper(II) sulfate solution of concentration 0.250 mol dm⁻³.

  (a) Calculate the mass of hydrated copper(II) sulfate required to prepare this solution. [2]

  (b) Describe, giving full practical details, how Frances should prepare the 250.0 cm³ of copper(II) sulfate solution. [6]

8 Elinor titrates a solution of potassium carbonate against a $0.2\,mol\,dm^{-3}$ hydrochloric acid solution to find the concentration of the potassium carbonate.

$$K_2CO_3 + 2HCl \longrightarrow 2KCl + CO_2 + H_2O$$

She uses methyl orange as an indicator. This turns from yellow in the potassium carbonate solution to pink when the potassium carbonate is neutralised by the hydrochloric acid.

She obtained the following results using $25.0\,cm^3$ samples of the potassium carbonate solution.

Titration	1	2	3	4
Final reading (cm³)	23.50	24.10	24.10	23.40
Initial reading (cm³)	0.40	0.15	0.90	0.25
Titre (cm³)				

(a) Calculate the mean titre that Elinor should use in her calculations. [2]

(b) Describe the practical steps that Elinor used to obtain a titration value. You should start by measuring $25.0\,cm^3$ of the potassium carbonate solution from a $250\,cm^3$ stock solution, with the acid already in the burette. [5]

(c) Elinor's value was slightly lower than the actual value. When asked why, she stated 'I did not add the acid drop by drop at the end and so overshot the end-point'.

State **two other** common sources of error in such experiments and explain why Elinor's statement cannot be correct. [4]

(Assume that all the equipment is clean and all the chemicals used are pure.)

(d) Rhodri carried out the titration using a $2.0\,mol\,dm^{-3}$ hydrochloric acid solution instead of a $0.2\,mol\,dm^{-3}$ solution. State if he is likely to have got a more accurate result for the concentration of the potassium carbonate solution. Justify your answer. [2]

Thermochemistry

p94

- Chemical reactions release energy to their surroundings (exothermic) or gain energy from their surroundings (endothermic)
- Energy can be changed from one form to another, it cannot be created or destroyed
- Enthalpy change is the name given to the energy exchange between a system and its surroundings under conditions of constant pressure
- Simple laboratory experiments can give estimates of the energy transferred during some reactions. Enthalpy changes may then be calculated
- Hess's Law is used to draw energy cycles to calculate the enthalpy change for reactions that cannot easily be measured directly

The wider impact of chemistry

p118

- The social, economic and environmental impact of chemical synthesis
- Factors associated with the production of energy
- Fossil and biomass fuels
- Carbon neutrality
- Nuclear, solar and hydrogen power; wind and water
- The role of green chemistry
- Opportunities to study how science works

Rates of reaction

p106

- Rates can be followed experimentally by gas collection, precipitation methods, colorimetry
- Rates can be calculated by dividing the change in concentration by the time taken for the change
- For a chemical change to occur molecules must collide with sufficient energy (activation energy)
- Increased concentration of molecules increases the frequency of collisions which increases the reaction rate
- Increased temperature increases the proportion of molecules with sufficient energy to react, which increases reaction rate
- A catalyst increases the reaction rate without being used up in the reaction. It provides an alternative route of lower activation energy

Organic compounds

p123

- Formulae can be shown in several different ways
- Nomenclature rules allow every organic compound to have a unique name that is recognised throughout the world by the scientific community
- Changing the structure of a molecule affects the physical properties
- Many compounds can show structural isomerism
- Reactions of organic compounds can be classified into different types according to the mechanism

Hydrocarbons p134

- In the use of hydrocarbons as fuels there are benefits and difficulties
- Bonds can be σ or π and the type of bond present affects the reactivity of the compound – compounds with π bonds are generally more reactive
- Alkanes undergo radical substitution reactions whilst alkenes undergo electrophilic addition reactions
- The presence of a π bond restricts rotation and allows the existence of *E-Z* isomers
- Alkenes undergo addition polymerisation to give a huge range of industrially and domestically important polymers

Alcohols and carboxylic acids p151

- Alcohols contain the functional group OH and carboxylic acids contain the functional group COOH
- The use of biofuels has advantages but also creates concerns
- Alcohols are classified as primary, secondary and tertiary and this affects what happens when oxidation is attempted
- Carboxylic acids show normal reactions as an acid with bases and carbonates but also react with alcohols to give esters

Halogenoalkanes p145

- Halogenoalkanes can take part in elimination reactions and in nucleophilic substitution reactions
- The rate of hydrolysis of halogenoalkanes is affected by the nature of the halogenoalkane used
- Halogenoalkanes have a wide variety of possible uses but there are serious environmental issues involved in their use
- Chemists consider bond strengths to try to identify alternatives to using CFCs

Instrumental analysis p160

- Instrumental techniques are now very widely used
- Each spectrum gives different information about the compound under investigation
- Mass spectra show the M_r of the compound, and the *m/z* ratio of fragments gives information about its structure
- IR spectra show the presence of particular bonds in a compound and hence the functional groups present
- ^{13}C NMR spectra show the number of different environments of the carbons present in the unknown
- ^1H NMR spectra show the number of different environments of the hydrogens present and also how many are present in each environment

Unit 2

2.1 Thermochemistry

All chemical reactions involve change. These energy changes are vital to us. Plants need the energy from the Sun for the production of carbohydrates by photosynthesis: we depend on the energy content of the food we eat. The kind of life we lead depends on harnessing energy from different sources.

There are many forms of energy but basically there are only two kinds of energy, kinetic energy and potential energy. Heat is a form of kinetic energy while the energy of chemical bonds is a form of potential energy. Thermochemistry is the study of the energy changes that accompany chemical reactions.

Topic contents

You should be able to demonstrate and apply knowledge and understanding of:

- Enthalpy change of reaction, enthalpy change of combustion and standard molar enthalpy change of formation.
- Hess's law and energy cycles.
- The concept of average bond enthalpies and how they are used to carry out simple calculations.
- How to calculate enthalpy changes.
- Simple procedures to determine enthalpy changes.

Temperature changes

Matter possesses energy in the form of kinetic energy and potential energy.

The kinetic energy of matter is the energy of motion at a molecular level. The potential energy of matter arises from the positions of the atoms relative to one another. Bond-breaking and bond-making involve changes in potential energy.

The sum of the kinetic energy of all the particles in a system and their potential energy is the internal energy of a system.

In the course of a chemical reaction, existing bonds are broken and new ones are made. This changes the chemical energy of atoms and energy is exchanged between the chemical system and the surroundings. Frequently this leads to heat being given out to or taken in from the surroundings.

Most chemical reactions release energy to their surroundings. This can be detected by a rise in the temperature of the reaction mixture and the surroundings. These reactions are known as **exothermic** reactions. Examples are:

- Acids with metals
- In hand warmers (oxidation of iron)
- Thermite reaction (aluminium and iron(III) oxide).

In some reactions the system absorbs energy from its surroundings in the form of heat. Such reactions are known as **endothermic** reactions. Examples are:

- Melting ice
- In cold packs (dissolving ammonium chloride in water)
- Thermal decomposition of group 2 carbonates.

Enthalpy changes

The amount of heat transferred in a given chemical reaction depends on the conditions under which the reaction occurs. Most chemical reactions in the laboratory take place under constant pressure. The total energy content of a system held at constant pressure is defined as its **enthalpy**, *H*.

Just like internal energy, enthalpy cannot be measured directly. However, an **enthalpy change**, ΔH, can easily be measured. Its units are joules, J, or kilojoules, kJ.

$$\Delta H = H_{products} - H_{reactants}$$

For exothermic changes (i.e a reaction that releases heat) heat is given out to the surroundings so $H_{products} < H_{reactants}$ and ΔH is negative.

For endothermic changes (i.e a reaction that absorbs heat) heat is taken in from the surroundings so $H_{products} > H_{reactants}$ and ΔH is positive.

Enthalpy changes can be represented as enthalpy profile diagrams.

YOU SHOULD KNOW ›››

››› that chemical reactions are accompanied by energy changes which result in temperature changes

››› what is meant by exothermic and endothermic reactions (prior GCSE knowledge)

Key Terms

An **exothermic reaction** is one that releases energy to the surroundings, there is a temperature rise and ΔH is negative.

An **endothermic reaction** is one that takes in energy from the surroundings, there is a temperature drop and ΔH is positive.

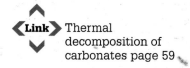

Link Thermal decomposition of carbonates page 59

Key Terms

Enthalpy, *H*, is the heat content of a system at constant pressure.

Enthalpy change, ΔH, is the heat added to a system at constant pressure.

Enthalpy profile diagram for an exothermic reaction (ΔH is negative):

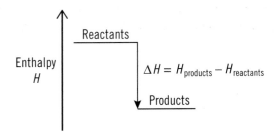

Enthalpy profile diagram for an endothermic reaction (ΔH is positive):

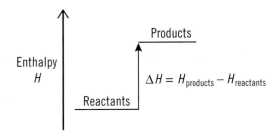

For example, when hydrogen and oxygen combine to form water:

$$H_2(g) + \tfrac{1}{2}O_2(g) \longrightarrow H_2O(g) \qquad \Delta H = -242\,kJ\,mol^{-1}$$

$$H_2(g) + \tfrac{1}{2}O_2(g) \longrightarrow H_2O(l) \qquad \Delta H = -286\,kJ\,mol^{-1}$$

this can be shown on an enthalpy diagram:

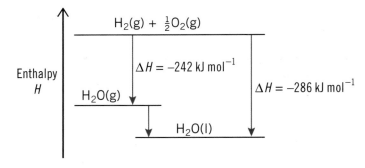

From this it can be seen that for

$$H_2O(g) \longrightarrow H_2O(l) \qquad \Delta H = -44\,kJ\,mol^{-1}$$

therefore it is an exothermic reaction.

Conservation of energy

In all the examples in this chapter we have assumed that all the energy that leaves a system enters the surroundings (and vice versa). In an exothermic reaction, energy is not being created, and it is not being destroyed in an endothermic reaction. The total energy of the whole system of reacting chemicals and surroundings is constant. This is an important principle and is called the **principle of conservation of energy.**

Standard conditions

Since enthalpy change for reactions depends on the conditions, for values to be compared standard enthalpy change is measured when fixed conditions are used. The conditions are:

- All substances in their standard states
- A temperature of 298 K (25 °C)
- A pressure of 1 atm (101 000 Pa).

The symbol for a standard enthalpy change is $\Delta H^{\ominus}$

YOU SHOULD KNOW ›››

››› what is meant by enthalpy change of reaction, enthalpy change of combustion and standard molar enthalpy change of formation

Standard enthalpy change of formation, $\Delta_f H$

This is the enthalpy change when one mole of a substance is formed from its constituent elements in their standard states under standard conditions.

For example, the standard enthalpy change of formation of carbon dioxide is represented as:

$$C(graphite) + O_2(g) \longrightarrow CO_2(g) \qquad \Delta_f H^{\ominus} = -394 \, kJ \, mol^{-1}$$

while the standard enthalpy change of formation of carbon monoxide is represented as:

$$C(graphite) + \tfrac{1}{2}O_2(g) \longrightarrow CO(g) \qquad \Delta_f H^{\ominus} = -111 \, kJ \, mol^{-1}$$

The 'per mole' refers to the formation of one mole of compound, not to the quantity of elements. Therefore the equation

$$2C(graphite) + O_2(g) \longrightarrow 2CO(g)$$

does not represent the enthalpy change of formation of carbon monoxide as two moles of carbon monoxide have formed.

If we are forming an element, such as $H_2(g)$, from the element $H_2(g)$, there is no chemical change. Therefore all elements in their standard state have a standard enthalpy change of formation of 0 kJ mol^{-1}.

> ▼ **Study point**
>
> Standard state of a substance is the substance in its pure form at 1 atm and the stated temperature (normally 298 K).

> ▼ **Study point**
>
> When we write thermochemical equations, we may need to use fractions like '$\tfrac{1}{2}O_2$', so that we have the correct number of moles involved in the change. In the example the standard enthalpy change of formation refers to the carbon monoxide, so to get a balanced equation we need half a mole of oxygen molecules.

Standard enthalpy change of combustion, $\Delta_c H$

This is the enthalpy change when one mole of a substance is completely combusted in oxygen under standard conditions.

For example, the standard enthalpy change of combustion of methane is given by:

$$CH_4(g) + 2O_2(g) \longrightarrow CO_2(g) + 2H_2O(l) \qquad \Delta_c H^{\ominus} = -891 \, kJ \, mol^{-1}$$

while that of hydrogen is given by:

$$H_2(g) + \tfrac{1}{2}O_2(g) \longrightarrow H_2O(l) \qquad \Delta_c H^{\ominus} = -286 \, kJ \, mol^{-1}$$

This time the 'per mole' refers to the substance being combusted not to the quantity of products formed.

> *Exam tip*
>
> If an element exists in more than one state, the standard state used should be the one that is most stable at 1 atm and 298 K. For carbon, graphite is more stable than diamond so in enthalpy equations, C(graphite) should be used rather than C(s).

> *Exam tip*
>
> You do not have to learn the definitions of these enthalpy changes but you will have to use them in calculating standard enthalpy changes and to write thermochemical equations.

Knowledge check

1

State what is meant by the standard enthalpy change of formation, $\Delta_f H^{\ominus}$.

Knowledge check

2

In the reaction:

$2H_2S(g) + 3O_2(g) \longrightarrow 2SO_2(g) + 2H_2O(l)$

Explain why the standard enthalpy change of formation for $O_2(g)$ is zero.

Enthalpy change of reaction, $\Delta_r H$

This is the enthalpy change in a reaction between the number of moles of reactants shown in the equation for the reaction.

There is no reason why a standard enthalpy change of reaction should be related to one mole of reactants. Therefore it is necessary to make clear which reaction equation is being used when an enthalpy change of reaction is being quoted.

For example, burning ethane in oxygen can be written as:

(i) $2C_2H_6(g) + 7O_2(g) \longrightarrow 4CO_2(g) + 6H_2O(l)$ $\Delta_r H = -3120\,kJ\,mol^{-1}$

(ii) $C_2H_6(g) + 3\frac{1}{2}O_2(g) \longrightarrow 2CO_2(g) + 3H_2O(l)$ $\Delta_r H = -1560\,kJ\,mol^{-1}$

Note that the value of $\Delta_r H$ in (i) is twice that of $\Delta_r H$ in (ii).

Calculating enthalpy change of reaction

The standard enthalpy change for a chemical reaction can be calculated from the standard enthalpy changes of formation of all reactants and products involved. The standard enthalpy of reaction, $\Delta_r H^{\ominus}$ is given by:

$$\Delta_r H = \Sigma\Delta_f H \,(products) - \Sigma\Delta_f H(reactants)\qquad(\Sigma \text{ stands for 'sum of'})$$

For example, calculate the standard enthalpy change of reaction for:

$$CS_2(l) + 4NOCl(g) \longrightarrow CCl_4(g) + 2SO_2(g) + 2N_2(g)$$

given the following

Compound	$CS_2(l)$	$NOCl(g)$	$CCl_4(g)$	$SO_2(g)$
$\Delta_f H^{\ominus} / kJ\,mol^{-1}$	88	53	−139	−296

$\Delta_r H = \Sigma\Delta_f H \,(products) - \Sigma\Delta_f H \,(reactants)$

$\quad = (-139 + 2(-296) + 2(0)) - (88 + 4(53))$ (remember that $\Delta_f H$ for any

$\quad = -731 - 300$ element in its standard state is 0)

$\quad = -1031\,kJ\,mol^{-1}$

Hess's Law

In the previous example, the standard enthalpy change of reaction depends only on the difference between the standard enthalpy of the reactants and the standard enthalpy of the products. However, in many chemical reactions the reactants may be able to change into the products by more than one route. Such reactions were studied by the Russian chemist Germain Hess and in 1840 he developed a chemical version of the principle of conservation of energy.

Hess's law states that the total enthalpy change for a reaction is independent of the route taken from the reactants to the products.

For example, consider the following enthalpy cycle showing two routes for converting reactants to products. The first is a direct route and the second an indirect route via the formation of an intermediate.

By Hess's Law the total enthalpy is independent of the route, so route 1 = route 2

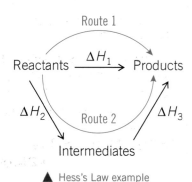

▲ Hess's Law example

i.e. $\Delta H_1 = \Delta H_2 + \Delta H_3$

If the sum of $\Delta H_2 + \Delta H_3$ was different from ΔH_1, it would be possible to create energy by making the products via the intermediate by one route and then converting back to the reactants by the other route. This would be contrary to the law of conservation of energy.

It follows that $\Delta H_1 = \Delta H_2 + \Delta H_3$, i.e. the standard enthalpy change is the same for the different routes. Note also that the enthalpy change of reaction from products to reactants is $-\Delta H_1$.

Hess's law enables us to calculate standard enthalpy changes of reactions that would be difficult to measure by using enthalpy cycles.

For example, forming nitrogen dioxide from nitric oxide:

$$NO(g) + \tfrac{1}{2}O_2(g) \longrightarrow NO_2(g)$$

The enthalpy changes of formation of the reactants and products are;

$\Delta_f H\ NO = 90.3\,kJ\,mol^{-1}$, $\Delta_f H\ NO_2 = 33.2\,kJ\,mol^{-1}$, $\Delta_f H\ O_2 = 0\,kJ\,mol^{-1}$.

The enthalpy cycle connecting the elements to the reactants and products is given by:

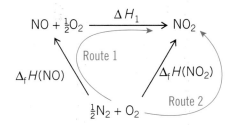

▲ Formation of NO_2

Since $\Delta_f H$ is given, the directions of the arrows go from the common elements to the reactants and products.

By Hess's law,   route 1 = route 2

$$\Delta H + 90.3 = 33.2$$

$$\Delta H = 33.2 - 90.3$$

$$\Delta H = -57.1\,kJ\,mol^{-1}$$

Or use the equation $\Delta_r H = \Sigma\Delta_f H\ (products) - \Sigma\Delta_f H\ (reactants)$

$$\Delta H = 33.2 - 90.3$$

$$\Delta H = -57.1\,kJ\,mol^{-1}$$

To find the enthalpy change of formation of ethane

$$2C(graphite) + 3H_2(g) \longrightarrow C_2H_6(g)$$

is impossible practically since burning carbon in hydrogen will produce a mixture of hydrocarbons. However, the enthalpy change of combustion of carbon, $\Delta_c H\ C = -394\,kJ\,mol^{-1}$, hydrogen, $\Delta_c H\ H_2 = -286\,kJ\,mol^{-1}$, and ethane, $\Delta_c H\ C_2H_6 = -1560\,kJ\,mol^{-1}$, can be measured accurately. Drawing an enthalpy cycle and using Hess's law can be used to calculate the enthalpy change of formation of ethane.

$$2C(s) + 3H_2(g) + 3\tfrac{1}{2}O_2(g) \xrightarrow{\ \Delta H_1\ } C_2H_6(g) + 3\tfrac{1}{2}O_2(g)$$

Route 1: $\Delta_c H\ (C_2H_6)$

Route 2: $2\,\Delta_c H(C) + 3\,\Delta_c H(H_2)$

$$2CO_2(s)\ +\ 3H_2O(l)$$

▲ Formation of ethane

Key Term

Hess's Law states that the total enthalpy change for a reaction is independent of the route taken from the reactants to the products.

▼ **Study point**

In an enthalpy cycle using $\Delta_c H$, the direction of the arrows goes from the reactants and products to the common combustion products.

In an enthalpy cycle using ΔH_f, the direction of the arrows goes from the common elements to the reactants and products.

Exam tip

Using enthalpy changes of combustion $\Delta H = \Sigma\Delta_c H(reactants) - \Sigma\Delta_c H(products)$

Using enthalpy changes of formation $\Delta H = \Sigma\Delta_f H(products) - \Sigma\Delta_f H(reactants)$

3

Knowledge check

Given the following enthalpy changes of formation

Substance	$H_2S(g)$	$S(s)$	$SO_2(g)$	$H_2O(l)$
$\Delta_f H^\ominus$/ kJ mol^{-1}	−20.2	0	−297	−286

Calculate the enthalpy change for the reaction

$$2H_2S(s) + SO_2(g) \longrightarrow 3S(s) + 2H_2O(l)$$

4

Since Δ_cH is given, the directions of the arrows go from the reactants and products to the common combustion products.

By Hess's law,   route 1 = route 2

$$\Delta H + (-1860) = (2(-394) + 3(-286))$$

$$\Delta H - 1860 = -1946$$

$$\Delta H = -1946 + 1860$$

$$\Delta H = -86 \, kJ \, mol^{-1}$$

Or use the equation:

$$\Delta H_r = \Sigma\Delta_cH \, (\text{reactants}) - \Sigma\Delta_cH \, (\text{products})$$

$$\Delta H = (2(-394) + 3(-286)) - (-1860)$$

$$\Delta H = -1946 + 1860$$

$$\Delta H = -86 \, kJ \, mol^{-1}$$

Bond enthalpies

Key Term

Bond enthalpy is the enthalpy required to break a covalent X — Y bond into X atoms and Y atoms, all in the gas phase.

Average bond enthalpy is the average value of the enthalpy required to break a given type of covalent bond in the molecules of a gaseous species.

A way of calculating the enthalpy changes of reactions involving covalent compounds is to consider the enthalpy associated with each covalent bond. The amount of energy needed to break a covalent bond is called the **bond enthalpy**. The values of bond enthalpies are always positive because breaking a bond requires energy.

For example, the bond enthalpy of HCl(g) is 431 kJ mol^{-1} and is the enthalpy change for the process, HCl(g) $\longrightarrow$ H(g) + Cl(g).

The actual value of the bond enthalpy for a particular bond depends on the structure of the rest of the molecule, so a C — C bond in ethane, C_2H_6, has a slightly different value to a C — C bond in pentane, C_5H_{12}. Although the bond enthalpy of a given bond is similar in a wide range of compounds, **average bond enthalpies** calculated using the values from many different compounds are used.

Calculations of enthalpy changes of reactions based on average bond enthalpy values will not be as accurate as results derived from experiments with specific molecules. However, they usually give an accurate enough indication of the standard enthalpy change of reaction.

Calculations using bond enthalpies

There are four steps involved:

Step 1   Draw out each molecule to show the bonds (if not already given).

Step 2   Calculate the energy required to break all the bonds in the reactants (endothermic, therefore a positive value).

Step 3   Calculate the energy released in forming all the bonds in the products (exothermic, therefore a negative value).

Step 4   Add together the enthalpy changes.

▼ Study point

Breaking bonds requires energy, therefore is endothermic (so bond enthalpy is always positive).

Making bonds releases energy so is exothermic so bond enthalpy is always negative.

Worked example

Using average bond enthalpies calculate the standard enthalpy change of reaction for the complete combustion of methane

$$CH_4(g) + 2O_2(g) \longrightarrow CO_2(g) + 2H_2O(g)$$

Bond	C—H	O=O	C=O	O—H
Average bond enthalpy/ kJ mol^{-1}	413	496	805	463

Step 1   Draw out each molecule

Step 2   Calculate the energy required to break the bonds (endothermic).

Bonds broken:

$$4(C—H) + 2(O=O) = (4 \times 413) + (2 \times 496) = 2644 \text{ kJ mol}^{-1}$$

Step 3   Calculate the energy released when bonds are made (exothermic).

Bonds formed:

$$2(C=O) + 4(O—H) = (2 \times -805) + (4 \times -463) = -3462 \text{ kJ mol}^{-1}$$

Step 4   Add together the energy changes

$$\Delta H = \Sigma(\text{bonds broken}) + \Sigma(\text{bonds formed})$$

$$\Delta H = 2644 + (-3462) = -818 \text{ kJ mol}^{-1}$$

Exam tip

When asked to calculate enthalpy change using bond enthalpies, always draw out each molecule so that you can see the bonds broken and bonds made.

5

Knowledge check

Using average bond enthalpies, calculate the enthalpy change for the reaction

$$2H_2(g) + O_2(g) \longrightarrow 2H_2O(g)$$

Measuring enthalpy changes

You cannot directly measure the heat content (enthalpy) of a system but you can measure the heat transferred to its surroundings. This process involves carrying out the chemical change in an insulated container called a calorimeter. The change in the temperature inside the calorimeter caused by the enthalpy change of the reaction can be measured with a thermometer.

If the temperature change is recorded and the mass and specific heat capacity of the contents of the calorimeter are known then the enthalpy change can be calculated.

The specific heat capacity is the heat required to raise the temperature of 1 g of a substance by 1 K. The value for water is 4.18 J g^{-1} K^{-1}.

The relationship between the temperature change, ΔT, and the amount of heat transferred, q, is given by the expression:

$$q = mc\Delta T$$

where m is the mass of the solution

and c is the specific heat capacity of the solution.

For the purposes of the calculations, we assume that all the heat is exchanged with the solution alone, the solution has the same specific heat capacity as water and the density of the solution is 1 g cm^{-3}.

▼ Study point

A calorimeter is any vessel used for determinations of heat changes. It is named after the old unit of heat, the calorie (1 cal = 4.18 J).

▼ Study point

In the expression $q = mc\Delta T$, you do not need to change °C into K, because temperature change is being used not actual temperatures.

The minus sign is used in the expression $\Delta H = -q/n$ because if there is an increase in temperature, the reaction is exothermic and ΔH is negative.

Knowledge check

6

A mug holds 250 cm³ of water. How much energy is needed to heat the water from 20 °C to 100 °C?

(Assume that the specific heat capacity for water is 4.18 J g⁻¹ K⁻¹.)

Therefore $q = m \times 4.18 \times \Delta T$ and the mass will be the same as the volume of the solution. To obtain the maximum temperature change allowances are made for heat lost (or gained) to the surroundings. Therefore, temperatures of the solution are taken for a short period before mixing and for some time after mixing. A graph of temperature against time is plotted and the maximum temperature is obtained by extrapolating the graph back to the mixing time.

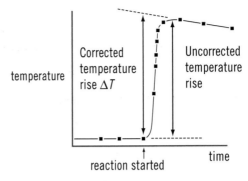

▲ Graph of temperature vs time

To calculate the enthalpy change of reaction per mole, we use the expression:

$$\Delta H = \frac{-q}{n}$$

Where n is the amount in moles that has reacted.

Worked examples

1 Displacement reactions

6g of zinc was added to 25.0 cm³ of 1.00 mol dm⁻³ copper(II) sulfate solution in a polystyrene cup. The temperature increased from 20.2 °C to 70.8 °C.

Calculate the enthalpy change for the reaction

$$Zn(s) + CuSO_4(aq) \longrightarrow ZnSO_4(aq) + Cu(s)$$

(Assume that the density of the solution is 1.00 g cm⁻³ and its specific heat capacity, c, is 4.18 J g⁻¹ K⁻¹.)

YOU SHOULD KNOW ›››

››› how to calculate enthalpy changes

Step 1   Calculate the amount of heat transferred in the experiment.

Only 25 cm³ of solution, therefore mass is 25 g

$$\Delta T = 70.8 - 20.2$$

$$q = mc\Delta T = 25 \times 4.18 \times 50.6 = 5288\,J$$

Step 2   Calculate the amount, in moles, of the reactants

$$\text{moles Zn} = \frac{m}{M} = \frac{6}{65.4} = 0.092$$

$$\text{moles CuSO}_4 = c \times \frac{V}{1000} = 1 \times 0.025 = 0.025$$

therefore $CuSO_4$ is not in excess and is used in the calculation

▼ Study point

Remember the m in step 1 is the mass of the solution that is changing the temperature not the mass of the solid.

Step 3   Calculate the molar enthalpy change

$$\Delta H = \frac{-q}{n} = \frac{-5288}{0.025} = -211\,520\,J\,mol^{-1}$$

$$= -211.5\,kJ\,mol^{-1}$$

2 Neutralisation reactions

25.0 cm³ of HCl(aq) are added to 25.0 cm³ of NaOH(aq) both of concentration 1.00 mol dm⁻³ in an insulated container. The acid and alkali are completely neutralised and the maximum temperature rise was calculated at 6.4 °C.

Exam tip

Show all your working. Each step of the calculation will be worth a mark.

Calculate the molar enthalpy change of neutralisation for the reaction

$$HCl(aq) + NaOH(aq) \longrightarrow NaCl(aq) + H_2O(l)$$

(Assume that the density of the solution is $1.00\,g\,cm^{-3}$ and its specific heat capacity, c, is $4.18\,J\,g^{-1}\,K^{-1}$.)

Step 1 Calculate the amount of heat transferred in the experiment.

Since $25.0\,cm^3$ of acid and alkali used, total volume of solution is $50.0\,cm^3$

$$q = 50 \times 4.18 \times 6.4 = 1338\,J$$

Step 2 Calculate the amount, in moles, of the reactants:

$$n = cv = 1.0 \times \frac{25.0}{1000} = 0.025$$

Step 3 Calculate the molar enthalpy change:

$$\Delta H = \frac{-q}{n} = \frac{-1338}{0.025} = -53\,520\,J\,mol^{-1}$$
$$= -53.5\,kJ\,mol^{-1}$$

3 Dissolving a solid to form an aqueous solution

$7.80\,g$ of ammonium nitrate, NH_4NO_3, were dissolved in $50.0\,cm^3$ of water in an insulated container. The temperature fell by $11.1\,°C$. Calculate the enthalpy of solution for the process:

$$NH_4NO_3(s) + aq \longrightarrow NH_4^+(aq) + NO_3^-(aq)$$

(Assume that the density of the solution is $1.00\,g\,cm^{-3}$ and its specific heat capacity, c, is $4.18\,J\,g^{-1}\,K^{-1}$.)

Step 1 Calculate the amount of heat transferred in the experiment:

$$q = mc\Delta T = 50 \times 4.18 \times (-11.1) = -2320\,J$$

Step 2 Calculate the amount, in moles, of the reactant:

$$n = \frac{m}{M} = \frac{7.80}{80.04} = 0.0975$$

Step 3 Calculate the molar enthalpy change:

$$\Delta H = \frac{-q}{n} = \frac{-(-2320)}{0.0975} = 23\,795\,J\,mol^{-1}$$
$$= 23.8\,kJ\,mol^{-1}$$

4 Combustion of alcohol

The combustion of $0.75\,g$ of ethanol raised the temperature of $250\,cm^3$ of water by $19.5\,°C$. Calculate the enthalpy change of combustion of ethanol, C_2H_5OH.

(The specific heat capacity, c, of water is $4.18\,J\,g^{-1}\,K^{-1}$.)

Step 1 Calculate the amount of heat transferred in the experiment:

$$q = mc\Delta T = 250 \times 4.18 \times 19.5 = 20\,378\,J$$

Step 2 Calculate the amount, in moles, of the reactant:

$$n = \frac{m}{M} = \frac{0.75}{46.06} = 0.0163$$

Step 3 Calculate the molar enthalpy change:

$$\Delta H = \frac{-q}{n} = \frac{-20\,378}{0.0163} = -1\,250\,184\,J\,mol^{-1}$$
$$= -1250\,kJ\,mol^{-1}$$

7

Knowledge check

$40.0\,cm^3$ of $HCl(aq)$ are added to $40.0\,cm^3$ of $NaOH(aq)$, both of concentration $0.80\,mol\,dm^{-3}$, in an insulated container. The acid and alkali are completely neutralised and the maximum temperature rise was calculated at $5.2\,°C$.

Calculate the molar enthalpy change of neutralisation for the reaction.

(The specific heat capacity, c, of water is $4.18\,J\,g^{-1}\,K^{-1}$.)

Experimental procedures

Combustion

Measurements of enthalpy changes of combustion are important as they help to compare the energy available from the oxidation of different flammable liquids that may be used as fuels.

The diagram below shows a simple method for obtaining an approximate value for the enthalpy of combustion of a fuel such as an alcohol.

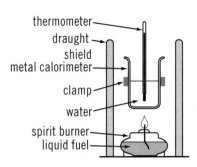

▲ Ethanol combustion

Since the mass of the solution is used in the expression to calculate ΔH, it has to be measured accurately. The number of moles is needed to calculate the molar enthalpy change, therefore the fuel also has to be measured accurately.

These are the main points to note in the practical work:

- Allow a suitable gap between the base of the metal container and the top of the spirit burner.
- Accurately measure the amount of water being added to the metal container.
- Use an accurate thermometer to measure the initial temperature of the water. When a steady value has been obtained record the temperature.
- Weigh the spirit burner containing the alcohol and record the initial mass.
- After lighting the wick, adjust the gap between the metal container and the spirit burner if necessary.
- Allow the alcohol to heat the water to a suitable temperature (an increase of about 20 °C is adequate – the smaller the increase, the greater the error on the thermometer).
- Extinguish the flame and record the final maximum temperature.
- Allow the spirit burner to cool thoroughly before reweighing and recording the final mass.

The value is much lower than the book value because:

- Some of the energy transferred from the burning alcohol is 'lost' in heating the apparatus and the surroundings.
- The alcohol is not completely combusted.
  (Soot (carbon) on the bottom of the calorimeter will show whether this has happened)

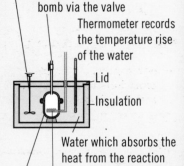
PRACTICAL CHECK

Determining an enthalpy change of combustion is a **specified practical task**.

To help increase the accuracy of the experiment:

The temperature can be measured every minute until 5 minutes after the temperature has reached a maximum value. A graph can be plotted and the maximum temperature obtained by extrapolating the graph back to the time when the burner was lit. This reduces the error caused by heat loss from the calorimeter to the surroundings.

A screen can be placed around the calorimeter to maximise the transfer of heat from the flame to the water.

The water in the calorimeter should be stirred continually to ensure an even temperature throughout.

Indirect determination of an enthalpy change

The simplest type of calorimeter is a coffee cup calorimeter. It can be used to measure changes that take place in aqueous solution.

The expanded polystyrene insulates the solution inside the cup so the amount of heat lost or absorbed by the cup during the experiment is negligible. (It can be placed inside another polystyrene cup or inside a beaker and lagged with cotton wool to improve insulation.)

Since the enthalpy change of many reactions, especially those involving solids (and gases), can be difficult to measure directly, solids are reacted with acids to form solutions. A coffee cup calorimeter can then be used.

All actual values obtained this way are lower than book values due to heat loss from the simple type of calorimeter used.

An example of indirect determination of enthalpy changes is the enthalpy change of reaction of magnesium oxide with carbon dioxide.

The enthalpy changes of reaction between the solid reactant (magnesium oxide) and acid and the solid product (magnesium carbonate) and acid are separately measured (hydrochloric acid is a suitable acid).The heat evolved is calculated, corrected for heat loss to the surroundings, scaled up to molar amounts of the solids involved and Hess's law is used to obtain the desired enthalpy change.

These are the main points to note in the practical work:

- Measure an appropriate volume of acid using a burette or pipette (the mass of acid is used in the expression to calculate ΔH) and place it in a polystyrene cup. This must be in excess (to ensure that all the solid reacts).

- Use an accurate thermometer to measure the initial temperature of the acid. When a steady value has been obtained record the temperature. (ΔT is used in the expression to calculate ΔH).

- Accurately weigh the solid, in powder form (to ensure as rapid a reaction as possible), in a suitable container (the amount, in moles, of the solid is used in the expression to calculate ΔH).

- Add all the solid to the cup, stir the mixture well (to ensure that the reaction is as rapid as possible and that all the solid is used) and start a stop-watch.

- Keep stirring with the thermometer and record the temperature regularly (about every 30 seconds). Stop recording the temperature when it has fallen for about 5 minutes.

- Re-weigh the weighing container to ensure that the correct mass of solid added is recorded.

- Plot a graph of temperature against time to calculate the maximum temperature the mixture might have reached. (This is essential to calculate the correct ΔT – see page 102).

- Calculate the amount of heat transferred ($q = mc\Delta T$).

- Calculate the enthalpy change for the reaction ($\Delta H = -q/n$).

- Repeat the procedure with the other solid.

- Use Hess's law to calculate the required enthalpy change.

This process is also suitable to find the enthalpy change of reaction for displacement reactions and for dissolving solids in water to find the enthalpy change of solution.

▲ Coffee cup calorimeter

YOU SHOULD KNOW ›››

››› how to carry out simple procedures to determine enthalpy changes

8

Knowledge check

When using a coffee cup calorimeter to determine the enthalpy change of a displacement reaction, explain why:

(a) The solid used is in powdered form.

(b) A graph of temperature against time is plotted and the graph extrapolated back to the mixing time.

PRACTICAL CHECK

Indirect determination of an enthalpy change of reaction is a **specified practical task**.

Exact details of indirect determinations will differ, but the procedures will remain the same.

Other suitable examples are:

Formation of magnesium oxide (adding magnesium and magnesium oxide separately to hydrochloric acid).

Formation of magnesium carbonate (adding magnesium and magnesium carbonate separately to hydrochloric acid).

Decomposition of sodium hydrogencarbonate (adding sodium hydrogencarbonate and sodium carbonate separately to hydrochloric acid).

Unit 2

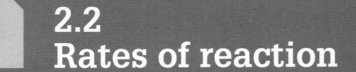

2.2
Rates of reaction

Many people are interested in knowing how to alter the rates of chemical reactions. Fertiliser manufacturers want to speed up the formation of ammonia. Car manufacturers want to slow down the rate at which iron rusts. Chemists seeking to manage environmental issues such as ozone depletion need to understand reaction rates.

This unit is concerned with measuring reaction rates in the laboratory, recognising what factors affect reaction rates and understanding how these factors affect reaction rates.

Topic contents

You should be able to demonstrate and apply knowledge and understanding of:

- How to calculate rates from experimental data and how to establish the relationship between reactant concentrations and rate.
- Collision theory in explaining the effects of changing conditions on reaction rate.
- The concepts of energy profiles and activation energy.
- The rapid increase in rate with temperature in terms of changes in the Boltzmann energy distribution curve.
- The characteristics of a catalyst
- How catalysts increase reaction rates by lowering activation energy.
- How colorimetry can be used in studies of some reaction rates.
- Measurements of reaction rate by gas collection and precipitation methods and by an 'iodine clock' reaction.

To many people speed is an essential component of a modern day life-style, whether it be in terms of travel, work or sport. People (and objects) move at a variety of speeds. Different chemical reactions also occur at different speeds, e.g. fireworks explode rapidly after ignition, metals react moderately quickly with acids, but rusting is a very slow process.

For chemical reactions, we use the term **rate** instead of speed. The study of rates of reaction is known as chemical kinetics. The **rate of a reaction** is found by measuring the amount of a reactant used up (or product formed) per unit of time. The amount is often expressed in terms of concentration.

For a reaction: rate $= \dfrac{\text{change in concentration}}{\text{time}}$ units: $\dfrac{\text{mol dm}^{-3}}{\text{s}} = \text{mol dm}^{-3}\text{s}^{-1}$

If another variable, such as mass or volume is measured, the rate can be expressed in corresponding units such as g s^{-1} or cm^3s^{-1}.

However, it is obvious from watching a reaction that the instantaneous rate changes as the reaction proceeds.

Usually for reactions:

- Rate is fastest at the start of a reaction since each reactant has its greatest concentration.
- Rate slows down as the reaction proceeds since the concentration of the reactants decreases.
- Rate becomes zero when the reaction stops, i.e. when one of the reactants has been used up.

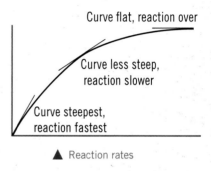

▲ Reaction rates

Calculating rates

We follow the rate of a reaction by measuring the concentration of a reactant (or product) over a period of time. The results obtained are plotted to give a graph. To find the initial rate, it is necessary to find the initial slope (gradient) of the line. At AS level the graph will always be a straight line to begin with, e.g.

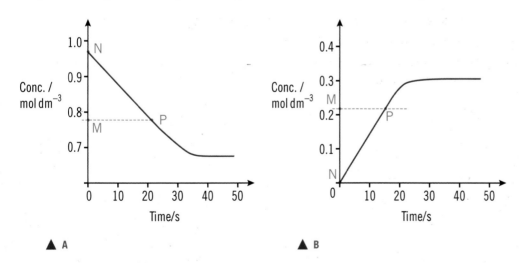

▲ A   ▲ B

To find the gradient: at any convenient point, P, on the straight line draw a horizontal line MP to the y axis and draw a vertical line from M to the beginning of the slope, N.

For graph A

$$\text{Rate} = \frac{\text{change in concentration}}{\text{time}} = \frac{MN}{MP} = \frac{(0.98 - 0.79)}{22} = \frac{0.19}{22} = 0.0086 \, \text{mol} \, \text{dm}^{-3} \text{s}^{-1}$$

For graph B

$$\text{Rate} = \frac{\text{change in concentration}}{\text{time}} = \frac{MN}{MP} = \frac{0.21}{16} = 0.013 \, \text{mol} \, \text{dm}^{-3} \text{s}^{-1}$$

To find out the relationship between initial rate and the initial concentrations of the reactants, a series of experiments, in which the concentration of only one reactant is changed at a time, must be performed.

$$\text{Since rate} = \frac{\text{change in concentration}}{\text{time}} \qquad \text{rate} \propto \frac{1}{\text{time}}$$

If a graph of $1/t$ is plotted against concentration, the relationship between reactant concentration and rate can be established.

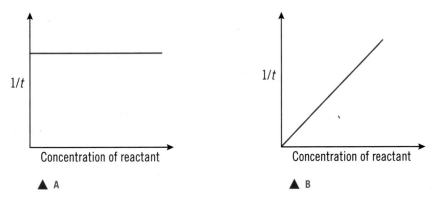

▲ A   ▲ B

Extra Help

You already know that

$$\text{speed} = \frac{\text{distance travelled}}{\text{time}}$$

So a car that travels 100 miles in two hours has an overall speed of

$$\frac{100 \text{ miles}}{2 \text{ hrs}} = 50 \text{ m.p.h.}$$

In a chemical reaction in which the concentration of a reactant changes from $1.20 \, \text{mol} \, \text{dm}^{-3}$ to $1.00 \, \text{mol} \, \text{dm}^{-3}$ in 30 seconds, the overall reaction rate is:

$$\text{Rate} = \frac{\text{change in concentration}}{\text{time}}$$

$$= \frac{(1.20 - 1.00) \, \text{mol} \, \text{dm}^{-3}}{30 \, \text{s}}$$

$$= 6.7 \times 10^{-3} \, \text{mol} \, \text{dm}^{-3} \text{s}^{-1}$$

In graph **A** although the concentration of the reactant has changed, the rate does not change, therefore the rate of reaction is independent of the concentration of that particular reactant.

In graph **B** as the concentration of the reactant increases, the rate also increases, therefore the rate of reaction is directly proportional to the concentration of that particular reactant.

If the results are tabulated, the results of the initial concentrations of reactants and rates must be compared.

For example, the table below gives the experimental data for the reaction between propanone, CH_3COCH_3, and iodine, I_2, carried out in dilute hydrochloric acid.

$$CH_3COCH_3(aq) + I_2(aq) \longrightarrow CH_3COCH_2I(aq) + HI(aq)$$

Knowledge check

Sodium thiosulfate reduces iodine to iodide ions. The rate of this reaction depends on the concentration of thiosulfate ions. At the beginning of a reaction between sodium thiosulfate and iodine, the concentration of thiosulfate ions was $0.100 \, \text{mol} \, \text{dm}^{-3}$. After three seconds the concentration had fallen to $0.088 \, \text{mol} \, \text{dm}^{-3}$. Calculate the initial rate of reaction.

Experiment	Initial concentrations / $\text{mol} \, \text{dm}^{-3}$			Initial rate
	$I_2(aq)$	$CH_3COCH_3(aq)$	$HCl(aq)$	/ $10^{-4} \, \text{mol} \, \text{dm}^{-3} \text{s}^{-1}$
1	0.0005	0.4	1.0	0.6
2	0.0010	0.4	1.0	0.6
3	0.0010	0.8	1.0	1.2

In experiments 1 and 2, only the concentration of iodine is changed and when it is doubled there is no change in the initial rate of reaction. Therefore the initial rate of reaction is independent of the initial concentration of aqueous iodine.

In experiments 2 and 3, only the concentration of propanone is changed and when it is doubled the initial rate of reaction also doubles. Therefore the initial rate of reaction is directly proportional to the initial concentration of aqueous propanone.

Collision theory

The explanation of rates of reaction is based on collision theory.

This says that for a reaction between two molecules to occur, an effective collision must take place, i.e. a collision that results in the formation of product molecules. The reaction rate is a measure of how frequently effective collisions occur.

Not all collisions between molecules result in reactions; however, the greater the number of collisions, the higher the chance that some of them will be effective. For a collision to be effective the molecules must collide in the correct orientation and with enough energy to react. Any factor that increases the rate of effective collisions will also increase the rate of the reaction.

YOU SHOULD KNOW ›››

››› how to use collision theory in explaining the effects of changing conditions on reaction rate

Factors that affect the rate of reaction

We know that a variety of factors can affect the position of equilibrium; similarly several factors may affect the rate of a chemical reaction.

1 Concentration (pressure for gases)

Increasing the concentration of reactants increases the rate of reaction, e.g. adding magnesium to concentrated acid will produce a more vigorous effervescence of hydrogen than will adding magnesium to dilute acid.

For reactions involving gases, increasing pressure will increase the rate of reaction, as pressure is proportional to concentration.

If there is an increase in concentration of reactants there are more molecules in a given volume. Distances between the molecules are reduced so there is an increase in the number of collisions per unit time. This means that there is a greater chance that the number of effective collisions increases, hence the rate of reaction increases.

For a gaseous reaction, increasing the pressure is the same as increasing the concentration.

▼ Study point

In collision theory, the term 'molecules' is used for simplicity. The species that collide could be molecules, atoms or ions.

▼ Study point

When explaining the effect of changing conditions on reaction rates always use collision theory in your answer. Bullet points can be useful.

2 Temperature

Increasing the temperature of reactants increases the rate of reaction, e.g. milk turns sour more quickly on a hot summer's day than in the cold winter.

If there is an increase in temperature there will be an increase in kinetic energy of the molecules and so they will move faster. This means that more molecules will have enough energy to react on collision, meaning that the reaction rate will increase.

3 Particle size

In a reaction involving a solid, breaking down the solid into smaller pieces increases the rate of reaction, e.g. powdered magnesium produces hydrogen far more quickly than magnesium ribbon when added to acid.

Reducing the particle size of a solid increases the surface area so again the molecules are closer together and there is an increase in the number of collisions per unit time leading to an increase in reaction rate.

4 Catalysts

A catalyst increases the rate of a chemical reaction without itself undergoing a permanent change, e.g. manganese(IV) oxide helps hydrogen peroxide decompose at room temperature.

5 Light

Some reactions are much more vigorous when carried out in bright light, e.g. photochlorination of methane.

10

Knowledge check

Explain why magnesium reacts faster when it is added to $2.0\,mol\,dm^{-3}$ hydrochloric acid than to $1.0\,mol\,dm^{-3}$ hydrochloric acid.

 Link Position of equilibrium pages 65–66

 Link Photochlorination of methane page 138

Energy profiles and distribution curves

Activation energy

Key Term

Activation energy is the minimum energy required to start a reaction by breaking of bonds.

Collision theory assumes that for a reaction to occur molecules must collide in a favourable orientation and they must collide with enough energy. This minimum kinetic energy is known as **activation energy, E_a,** and it varies from one reaction to another. If the reactants collide with an energy at least equal to the activation energy, the collision is successful and products will form.

The activation may be shown on diagrams called energy profiles. These compare the enthalpy of the reactants with the enthalpy of the products:

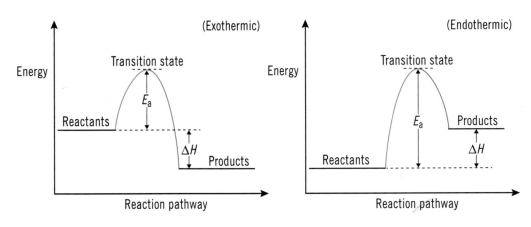

▲ Energy profile diagrams

These diagrams illustrate the role of activation energy as an energy barrier that must be overcome by reactants before they can form products.

In the first part of the energy profiles (coloured green in the diagrams), the reactant molecules are coming together and breaking apart. Separating atoms in the reactant molecules requires bonds to be broken, so energy is absorbed.

In the second part of the energy profile (coloured red in the diagrams), the product molecules are forming and moving apart. Producing product molecules involves forming bonds, so energy is released.

Therefore, even if a reaction is exothermic an input of energy is required to break bonds to start the reaction. Once the reaction has started, enough heat energy is produced to keep the reaction going.

The difference between the energy of the reactants and products is the enthalpy change of reaction. As seen from the profile, for an exothermic reaction the products have a lower energy than the reactants and so heat is given out. For an endothermic reaction the products have a higher energy than the reactants and heat is taken in from the surroundings.

The energy profile on the next page shows the implication of this for a reversible reaction:

Knowledge check

Calculate the enthalpy change for the reaction

$$2HI(g) \rightleftharpoons H_2(g) + I_2(g)$$

given that the activation energy for the forward reaction is $195\,kJ\,mol^{-1}$ and the activation energy for the backward reaction is $247\,kJ\,mol^{-1}$.

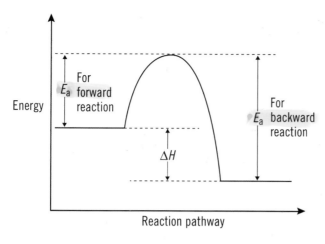

Energy

For
E_a forward
reaction

For
E_a backward
reaction

ΔH

Reaction pathway

▲ Energy profile for a reversible reaction

 Enthalpy changes
page 95

If the reversible reaction is exothermic in one direction, it will be endothermic in the reverse direction. Also the enthalpy change of reaction is given by:

$$\Delta H = E_a f - E_a b$$

where $E_a f$ and $E_a b$ are the activation energies of the forward and reverse reactions respectively.

For an exothermic reaction $E_a f < E_a b$ and ΔH is negative.

For an endothermic reaction $E_a f > E_a b$ and ΔH is positive.

The Boltzmann distribution

The molecules in a gas are moving constantly at different speeds so the energy of each molecule varies greatly. Since the molecules collide frequently, a molecule that has been knocked can move quicker with greater energy than before while the molecule that caused the collision will slow down and have hardly any energy.

The Boltzmann energy distribution curve shows the distribution of molecular energies in a gas. The energies of a few molecules are almost zero. Most molecules have an energy around an average value. Only a minority have values that equal or exceed the activation energy.

YOU SHOULD KNOW ›››

››› the rapid increase in rate with temperature in terms of changes in the Boltzmann energy distribution curve

Exam tip

When explaining the effect of temperature on reaction rate you must include a reference to activation energy in your answer.

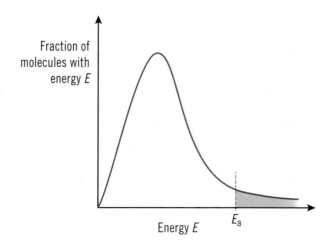

Fraction of
molecules with
energy E

Energy E E_a

▲ Diagram of Boltzmann distribution

At higher temperatures the average molecular energy will increase. Some molecules will still be almost motionless but at any one time many more molecules will have a higher energy. At two different temperatures, T_1 and T_2, where $T_2 > T_1$, the Boltzmann distribution curve shows why reaction rate is so dependent on temperature.

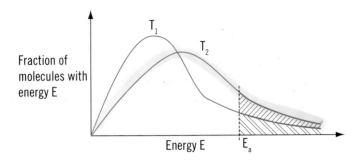

▲ Boltzmann distribution for two temperatures

- The curves do not touch the energy axis.
- The areas under the two curves are equal and are proportional to the total number of molecules in the sample.
- At the higher temperature, T_2, the distribution flattens, the peak moves to the right (higher energy) with a lower height.
- At the higher temperature, T_2, the mean energy of the molecules increases. There is a wider spread of values.
- Only the molecules with an energy equal to or greater than the activation energy, E_a, are able to react.
- At the higher temperature, T_2, many more molecules have sufficient energy to react and so the rate increases significantly.

Knowledge check

In the reaction between magnesium and hydrochloric acid, explain why raising the temperature of the acid increases the rate of reaction.

Catalysts

Many reactions have high activation energies. This means that in order to convert reactants to products at a reasonable rate, the reaction mixture needs to be maintained at a very high temperature. However, this difficulty may be overcome by the use of **catalysts**.

A catalyst is a substance that increases the rate of a chemical reaction without being used up in the process. A catalyst does take part in the reaction but can be recovered at the end of the reaction unchanged.

Catalysts work by providing a different reaction pathway for the reaction. The reaction rate increases because the new pathway has a lower activation energy than that of the uncatalysed reaction. This can be shown in an energy profile diagram:

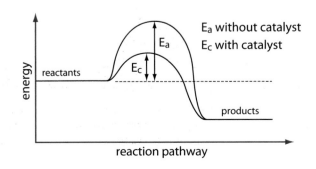

▲ Energy profile diagram for catalyst

Because the activation energy of the catalysed reaction is lower, at the same temperature a greater proportion of colliding molecules will achieve the minimum energy needed to react. This can be shown on an energy distribution curve diagram:

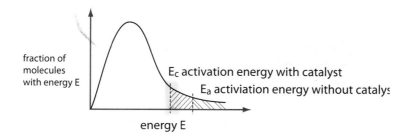

▲ Energy distribution curve for catalyst

For a reversible reaction, a catalyst increases the rate of the forward and back reactions by the same amount therefore it does not affect the position of equilibrium, but the position of equilibrium is reached more quickly.

There are two classes of catalysts: heterogeneous and homogeneous.

Homogeneous catalysts

A **homogeneous catalyst** is in the same phase as the reactants. They take an active part in a reaction rather than being an inactive spectator. Homogeneous catalysis typically involves liquid mixtures or substances in solution.

Examples are:

- Concentrated sulfuric acid in the formation of an ester from a carboxylic acid and an alcohol.
- Aqueous iron(II) ions, $Fe^{2+}(aq)$, in the oxidation of iodide ions, $I^-(aq)$, by peroxodisulfate(VI) ions, $S_2O_8^{2-}(aq)$.

Heterogeneous catalysts

A **heterogeneous catalyst** is in a different phase from the reactants.

Many heterogeneous catalysts are d-block transition metals. Gases are adsorbed on to the metal surface and react and the products desorb from the surface. The larger the surface area, the better the catalyst works.

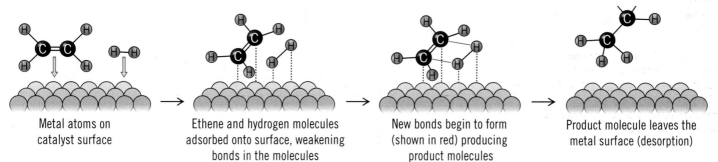

| Metal atoms on catalyst surface | Ethene and hydrogen molecules adsorbed onto surface, weakening bonds in the molecules | New bonds begin to form (shown in red) producing product molecules | Product molecule leaves the metal surface (desorption) |

▲ Catalyst mechanism

> **▼ Study point**
>
> A catalyst does not appear as a reactant in the overall equation of a reaction.

> **Key Terms**
>
> A **homogeneous catalyst** is in the same phase as the reactants.
>
> A **heterogeneous catalyst** is in a different phase from the reactants.

⟨**Link**⟩ Hydrogenation of alkenes page 141

Autocatalysis occurs when one of the products of the reaction is a catalyst for the reaction. Such a reaction starts slowly at the uncatalysed rate. As the concentration of the product (which is also the catalyst) builds up, the reaction speeds up to the catalysed rate. This leads to an odd-looking rate curve.

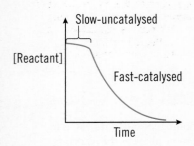

▲ Rate curve of autocatalysis

13

Knowledge check

State what is meant by a catalyst and explain how catalysts work.

Link Green chemistry pages 121–122

YOU SHOULD KNOW ›››

››› how colorimetry can be used in studies of some reaction rates

Many important industrial processes involve the use of catalysts:

- Ziegler–Natta catalysts in production of high density poly(ethylene).
- Iron in the Haber process for ammonia production.
- Vanadium(V) oxide in the contact process within sulfuric acid manufacture.
- Nickel in the hydrogenation of unsaturated oils in the production of margarine.

Industry relies on catalysts to reduce costs. Since a catalyst speeds up a process by lowering the activation energy of the reaction, less energy is required for the molecules to react, and this saves energy costs. Much of this energy is taken from electricity supplies or by burning fossil fuel so a catalyst also has benefits for the environment. If less fossil fuel is burned less carbon dioxide will be released during energy production.

Many plastics are formed under high pressures. If a catalyst can be found to enable the reaction to give a good yield at low pressure, the industrial plant will not have to withstand high pressure and less robust materials can be used in the construction, thus saving money.

Enzymes

Enzymes are biological catalysts. They are complex globular proteins which act as homogeneous catalysts in living systems. They usually catalyse specific reactions and work best close to body temperature. Enzyme activity is affected by temperature (it increases until the protein denatures) and pH (different enzymes have differing optimum pH levels). However, enzymes are very effective catalysts, giving a greater increase in reaction rates than inorganic catalysts, e.g. one catalase molecule decomposes 50 000 molecules of hydrogen peroxide per second.

Enzymes are used in a variety of important industrial processes such as food and drink production and the manufacturing of detergents and cleaners. Some examples are:

- Rennin in the dairy industry.
- Yeast and amylase in the brewing industry.
- Lipase and protease in washing powders and detergents.

Some of the benefits are:

- Lower temperatures and pressures can be used, saving energy and costs.
- They operate in mild conditions and do not harm fabrics or food.
- They are biodegradable. Disposing of waste enzymes is no problem.
- They often allow reactions to take place which form pure products, with no side reactions removing the need for complex separation techniques.

Studying rates of reaction

To measure the rate of a chemical reaction we need to find a physical or chemical quantity which varies with time. These are some of the main methods (all carried out at constant temperature).

- **Change in gas volume**

  In a reaction in which gas is formed, the volume of the gas can be recorded using a gas syringe at various times.

  e.g. $Mg(s) + 2HCl(aq) \longrightarrow MgCl_2(aq) + H_2(g)$

  or $2H_2O_2(aq) \longrightarrow 2H_2O(l) + O_2(g)$

- **Change in gas pressure**

  Some reactions between gases involve a change in the number of moles of gas. The change in pressure (at constant volume) at various times can be followed using a manometer.

  e.g. $PCl_5(g) \longrightarrow PCl_3(g) + Cl_2(g)$

- **Change in mass**

  If a gas forms in a reaction and is allowed to escape, the change in mass at various times can be followed using weighing scales.

  e.g. $CaCO_3(s) + 2HCl(aq) \longrightarrow CaCl_2(aq) + H_2O(l) + CO_2(g)$

- **Colorimetry**

  Some reaction mixtures show a steady change of colour as the reaction proceeds. The concentration of the substance changing colour can be monitored using a colorimeter.

  A colorimeter consists of a light source with filters to select a colour of light which is absorbed by the sample. The light passes through the sample and onto a detector or photocell. The photocell which is connected to a computer develops an electrical signal proportional to the intensity of the light. The colorimeter is calibrated with solutions of known concentration to establish the relationship between its readings and the concentration of the species being observed.

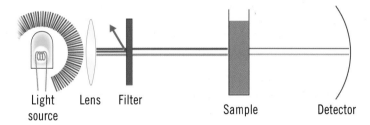

Light   Lens   Filter   Sample   Detector
source

▲ Colorimeter

For example, it could be used for the investigation of iodine with propanone $CH_3COCH_3(aq) + I_2(aq) \longrightarrow CH_3COCH_2I(aq) + HI(aq)$ since iodine (brown) is the only coloured species in the reaction.

Practical activity

A good way of showing how rate changes during a chemical reaction as well as illustrating how changing concentration, temperature, particle size or catalysts can affect a chemical reaction is the **gas collection method**.

For example, reacting magnesium with an acid. A typical method would be:

- Set up the apparatus as shown (ensuring that one reactant is in excess).
- Start the reaction by shaking the magnesium into the acid and start a stopwatch.
- Measure the amount of hydrogen given off at constant intervals (e.g. 30 s).
- Stop the watch when hydrogen is no longer being produced.
- Repeat the experiment with different concentrations of acid / temperature of acid / particle size of magnesium ensuring that all other factors are kept constant.
- Draw a graph of your results.

Another main method to study reaction rate is sampling and titration.

The reaction mixture is sampled by pipette at suitable time intervals and the concentration of one of the reactants (or products) found by a suitable titration. The reaction in each sample taken is slowed down significantly (e.g. by diluting in ice-cold water) so that titration can be carried out at leisure.

e.g. $CH_3CO_2C_2H_5(aq) + H_2O(l) \longrightarrow$
$$CH_3CO_2H(aq) + C_2H_5OH(aq)$$

or $CH_3COCH_3(aq) + I_2(aq) \xrightarrow[\text{catalyst}]{H^+}$
$$CH_3COCH_2I(aq) + HI(aq)$$

In the second example, the sampled mixture can be run into a flask containing sodium hydrogencarbonate. This neutralises the acid catalyst, slowing down the reaction considerably. The iodine remaining can be titrated against sodium thiosulfate solution to calculate its concentration.

Why is colorimetry a better method than sampling and titration for following the rate of reaction for the reaction between propanone and iodine using an acid catalyst?

14

Knowledge check

Suggest a method for following the rate of the following reaction:

$2FeCl_3 + 3Na_2CO_3 + 3H_2O$
$$\longrightarrow 6NaCl + 2Fe(OH)_3 + 3CO_2$$

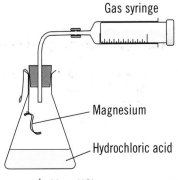

Gas syringe

Magnesium

Hydrochloric acid

▲ Mg + HCl

PRACTICAL CHECK

Investigation of a rate of reaction by a gas collection method is a **specified practical task.**

Other suitable examples are:

Addition of calcium carbonate to hydrochloric acid. (Sulfuric acid is not suitable.)

Decomposition of hydrogen peroxide. (Different catalysts can be compared.)

Instead of using a gas syringe, the gas can be collected over water using an inverted burette.

15

Knowledge check

In the reaction between magnesium and hydrochloric acid, why should you not start the reaction by dropping the magnesium into the acid and replacing the bung?

PRACTICAL CHECK

Investigation of a rate of reaction by studying an 'iodine clock' reaction is a **specified practical task.**

In the reaction between iodide ions and hydrogen peroxide in acid solution, a large batch containing acid, iodide, thiosulfate and starch in the correct proportions can be made up if preferred. When investigating the effect of peroxide concentration, 36.0 cm³ of this mixture can be added to a conical flask, using a burette, each time a different concentration of peroxide solution is used.

Another suitable example is:

Oxidation of iodide ions by peroxodisulfate ions to form iodine.

Add a fixed volume of aqueous thiosulfate ions and a few drops of starch solution to the reaction mixture. The thiosulfate rapidly reduces the iodine formed. When all the thiosulfate has reacted, free iodine quickly forms a blue complex with the starch.

Analysis of a mixture during the course of a reaction is not always possible or required. To compare rates of reaction under different conditions, a number of experiments may be set up in which initial concentrations of reactants are known and the time taken for each experiment is recorded.

Two examples of this are 'iodine clock' and precipitation reactions.

Iodine-clock reactions

Iodide ions can be oxidised to iodine at a measurable rate. Iodine gives a strongly coloured blue complex with starch solution but if a given amount of thiosulfate ion – with which iodine reacts very rapidly – is added, no blue colour will appear until enough iodine has been formed to react with all the thiosulfate. The time taken for this to occur thus acts as a 'clock' to measure the rate of iodide ions being oxidised.

Oxidising iodide ions by hydrogen peroxide in acid solution is a suitable example of an 'iodine clock' reaction:

$$H_2O_2(aq) + 2H^+(aq) + 2I^-(aq) \xrightarrow{\text{slow}} 2H_2O(l) + I_2(aq)$$

$$I_2(aq) + 2S_2O_3^{2-}(aq) \xrightarrow{\text{fast}} 2I^-(aq) + S_4O_6^{2-}(aq)$$

By varying the concentrations of the reactants one at a time and measuring the rate, the dependence of rate on concentration for any reactant may be found. A trial run to find what range of concentrations will be suitable should be performed first. The temperature must be kept constant, since rates vary rapidly with changes in temperature.

A typical method for this reaction would be:

- For the trial, add 10.0 cm³ H_2SO_4 (1 mol dm⁻³), 10.0 cm³ $Na_2S_2O_3$ (0.005 mol dm⁻³), 15.0 cm³ KI (0.1 mol dm⁻³) and 9.0 cm³ deionised water from burettes into a conical flask.

- Add 1 cm³ of starch solution.

- Measure 5.0 cm³ H_2O_2 (0.1 mol dm⁻³) from a burette into a test-tube.

- Rapidly pour the H_2O_2 into the flask, simultaneously start a stopwatch and mix thoroughly.

- When the blue colour appears stop the watch.

- Repeat using 5 different concentrations of peroxide, ensuring that the total volume of the mixture is 50 cm³.

- The concentration of peroxide should vary by at least threefold to ensure a good spread of results.

Now rate $\propto \dfrac{1}{\text{time}}$ and since total volume is constant in each case, $[H_2O_2] \propto$ volume of peroxide used in each run. Plotting a graph of $\dfrac{1}{\text{time}}$ against volume of peroxide will give the relationship between $[H_2O_2]$ and rate.

A similar procedure can be used varying the concentration of the potassium iodide to find the effect of $[I^-]$ on reaction rate. However, the peroxide solution must be added last every time.

Precipitation reactions

It is not only colour changes in solutions that can be used to follow rates of reaction. Sometimes colourless solutions become more and more cloudy as a precipitate forms and the time taken for a precipitate to form can be used to measure the rate of a reaction.

An example of this is the reaction of thiosulfate ions in acid solution.

$$S_2O_3^{2-}(aq) + 2H^+(aq) \longrightarrow S(s) + H_2O(l) + SO_2(g)$$

The reaction is easy to follow since one sulfur atom is formed for each thiosulfate ion reacting and the sulfur makes the reacting solution more cloudy as its concentration increases. By placing the reaction vessel over a black cross (which will disappear from view when a fixed amount of reaction has produced a fixed amount of sulfur) the rates of reaction of solutions of differing concentrations can be compared and the effects of changing concentrations on the rate found.

This is because the time taken for this fixed amount of reaction in all runs is inversely proportional to the rate of reaction, i.e. if the reaction is fast the cross will disappear quickly and vice versa.

A trial run to find what range of concentrations will be suitable should be performed first. The temperature must be kept constant, since rates vary rapidly with changes in temperature.

A typical method for this reaction would be:

- For the trial add 6.0 cm³ $Na_2S_2O_3$ (1 mol dm⁻³) and 4.0 cm³ water into a conical flask.
- Add 10.0 cm³ HNO_3 (0.1 mol dm⁻³) from a burette into a test tube.
- Rapidly pour the acid into the flask, simultaneously start a stopwatch and mix thoroughly.
- Place the flask over a black cross.
- Stop the watch the moment the black cross can no longer be seen.
- Repeat the procedure with three different concentrations of $Na_2S_2O_3$ keeping the volume of acid at 10.0 cm³ and the total volume of the mixture at 20.0 cm³. (Ensure that there is a good spread of concentration.)
- Repeat the procedure with three different concentrations of HNO_3 keeping the volume of thiosulfate at 6.0 cm³ and the total volume of the mixture at 20.0 cm³. (Ensure that there is a good spread of concentration.)

Since rate $\propto \dfrac{1}{\text{time}}$ plotting a graph of $\dfrac{1}{\text{time}}$ against concentration of thiosulfate at constant acid concentration will give the relationship between $[S_2O_3^{2-}]$ and rate.

Plotting a graph of $\dfrac{1}{\text{time}}$ against concentration of acid at constant thiosulfate concentration will give the relationship between $[HNO_3]$ and rate.

YOU SHOULD KNOW ›››

››› how to measure reaction rate by gas collection and precipitation methods and by an 'iodine clock' reaction

Knowledge check

When rates are measured in an iodine clock reaction, state why it is essential that:

(a) All the runs (experiments) are made with the same total volume of liquid.

(b) The temperature does not change during the series of runs.

Unit 2

2.3
The wider impact of chemistry

The large-scale production of chemicals and combustion of fossil fuels is causing serious problems especially in the rise of CO_2 levels and its effect on the climate. Efforts are being made to use less fossil fuels or to use alternative fuels and to use fuels more efficiently.

Green chemistry includes this in its attempt to prevent waste, use safer and more efficient methods, increase atom economy and reduce pollution.

Topic contents

You should be able to demonstrate and apply knowledge and understanding of the following:

- The serious problems facing the Earth and its inhabitants.
- Excessive dependence on fossil fuels has caused a large increase in atmospheric CO_2 levels and probable climate change.
- Fossil fuels are non-renewable and will eventually become exhausted.
- Carbon neutrality and how it may be achieved.
- Nuclear and solar power and an economic hydrogen fuel are possible alternatives.
- The role of green chemistry in both energy efficiency and in the waste-free, safe, high efficiency production of needed products using renewable raw materials and avoiding pollution and the use of unnecessary solvents.

(a) The social, economic and environmental impact of chemical synthesis and the production of energy

Both the chemical industry and obtaining the materials for chemical processes are major sources of employment and many towns have grown up around such centres. Industries have been sited near sources of raw materials such as iron ore, coal, water and good transport. There is now a greater realisation of the importance of providing a healthy and safe environment for the workers and their families living near the factories.

▲ New Lanark Mill – an early factory with housing for workers. Almost the first example of real consideration for factory workers and families.

Economically the requirements today are to produce good quality products efficiently and safely with a well-paid and fulfilled workforce.

▲ Wind power

Key Term

Carbon neutrality means that a chemical process such as fuel combustion does not lead to an overall increase in CO_2 levels. Although the combustion does generate CO_2 this is offset by the fact that the fuel, such as biomass or sugar cane, has absorbed the same amount of CO_2 in being made by photosynthesis as in the equation above.

17

Knowledge check

Which of the following combustion fuels is carbon neutral?

Coal – Oil – Methane – Hydrogen – Liquefied petroleum gas (LPG)

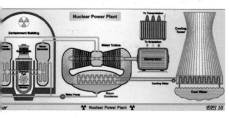

▲ Nuclear power

▲ Solar power

The energy problem

Energy production is a major issue affecting not only the chemical industry but all aspects of modern life and serious problems of supply lie ahead in that the finite and non-renewable resources of fossil fuels (coal, oil and gas) are being rapidly depleted.

Huge and increasing amounts of energy are needed by the world's expanding and developing population for industry, transport, electricity generation and domestic heating. The main sources are the non-renewable fossil fuels above, renewable sources including wind and water power in its various forms, the combustion or bacterial digestion of biomass (wood, sugar cane, animal wastes), geothermal energy, and nuclear and solar energy.

Hydrogen is spoken of as a future fuel but does not exist naturally on Earth and must be prepared at a cost of energy.

(i) Fossil and biomass fuels

All fossil and biomass fuels are derived from solar energy acting on plants and micro-organisms coal, oil and gas over hundreds of millions of years and biomass at the present day. The rate of usage of fossil fuels, whether obtained by mining, drilling, tar sands or fracking, is perhaps one hundred times their formation rate so that these will run out at some future date.

The essential processes are shown in the simple equation:

$$\underset{\text{sugar}}{CH_2O} + O_2 \underset{\text{photosynthesis}}{\overset{\text{combustion}}{\rightleftharpoons}} CO_2 + H_2O$$

(ii) Carbon neutrality

Underlying the energy problem is the knowledge that a huge amount of carbon dioxide is being passed into the atmosphere through fuel combustion and other industrial processes, such as making concrete. Its concentration has risen from around 300 to 400ppm (parts per million) in the last 100 years, and there seems little doubt that this has led to global warming through an increase in the greenhouse effect. The consequences of this warming may be extremely serious.

Carbon neutrality is achieved with combustible fuel sources by replanting trees and sugar cane, etc., to match those being burnt, so that the CO_2 consumed in their photosynthesis equals or exceeds that generated by combustion.

Leaving aside the purely physical processes, such as the use of wind and water, two other important energy sources that do not generate CO_2 are of chemical interest namely nuclear and solar energy which generate electricity directly.

(iii) Nuclear power

Around 20% of the electricity needed may be produced in this way by the neutron-aided fission of uranium in which less than 1% of mass is converted to energy through the equation $E = mc^2$. The method has worked for some fifty years but has drawbacks with regard to radioactive emissions and the safe disposal of radioactive wastes.

A better method on paper is in the fusion of hydrogen to form helium, as in the Sun, that converts a higher percentage of mass into energy with fewer problems of radioactive products. Unfortunately, despite several decades of research, it has not yet been possible to solve the associated technical problems.

(iv) Solar power

Solar power based on silicon or gallium arsenide semiconductor panels can achieve efficiencies of over 10% and is rapidly growing in use. In the UK average outputs of 0.5 kW per square metre may be obtained.

(v) Efficient use of energy

Much research is carried out towards making more efficient use of energy, especially in engines and considerable savings are being made. One of the main aims of green chemistry in (b) below is to use less energy in industrial processes such as through working at lower temperatures using effective catalysts.

The chemical industry

Efficient production of chemicals and use of energy requires an understanding of the basic principles covered in Units 1.7, 2.1 and 2.2.

This topic requires you to apply these principles to situations and problems encountered in the production of chemicals and energy. You may be supplied with data relevant to the situation, process or problem which you may not have met before and will be marked on your ability to analyse and evaluate the situation or problem, usually by answering a series of questions on the topic. Calculations may be required and clearly a basic understanding of equilibrium, energetics and kinetics will be important in tackling many of the questions. There are no specific learning outcomes as such, what is needed is some practice and experience in applying the relevant sections of Units 1.7, 2.1 and 2.2 to the particular question.

(b) Understand the role of green chemistry in improving sustainability in all aspects of developments

The aim of green chemistry is to make the chemicals and products that we need with as little impact on the environment as possible. This means:

Renewable raw materials
Using renewable raw materials such as plant-based compounds whenever possible.

Saving energy
Using as little energy as possible and getting this from renewable sources such as biomass, solar, wind and water rather than from finite fossil fuels such as oil, gas and coal.

High atom economy
Using methods having high atom economy so that a high percentage of the mass of reactants ends up in the product giving little waste (see page 42 Atom economy).

Catalysts
Developing better catalysts and biocatalysts, such as enzymes, to carry out reactions at lower temperatures and pressures to save energy and avoiding high-pressure plants (see page 112).

Toxic materials and solvents
Avoid using toxic materials if possible and ensure that no undesirable co-products or by-products are released into the environment.

 **Link** Fuels page 135

▲ Jet engine

Photoelectrochemical cells and the hydrogen economy.

Plants are only about 5% efficient in capturing sunlight. Hydrogen is a carbon-neutral fuel but requires energy to make.

Recent research uses sunlight to split water into hydrogen and oxygen in an electrochemical cell but needs much work to increase reliability and efficiency. The latter now exceeds 10%.

HOW SCIENCE WORKS

The above is a good example of this.

Aim – produce large amounts of cheap hydrogen fuel.

Principle – electrolyse water to hydrogen and oxygen in a cell.

Method – find a way of getting sunlight to do this.

Develop – improve the efficiency, reliability and practicality of the method to work on an industrial scale.

Avoid using solvents, especially volatile organic compounds (VOCs) that are bad for the environment. Diesel engines liberate toxic nitrogen oxides (NO_x) into the atmosphere. Addition of ammonia converts these into harmless nitrogen. (See bullet point below.)

Biodegradable products

Make products that are biodegradable at the end of their useful lives, where possible.

Summary of green principles

1 Prevent waste
2 Increase atom economy
3 Use safer methods, chemicals and solvents
4 Increase energy efficiency
5 Use renewable raw materials (feedstocks)
6 Use catalysts (vs. stoichiometric reactions)
7 Prevent pollution and accidents
8 Design for biodegradation.

Some examples involving the principles above:

- Anaerobic digestion – methanogenic bacteria act on manures and vegetable wastes in an oxygen-free atmosphere to produce biogas containing 70% methane for use in electricity generation and leave a fertilizer as a residue.

- Feedstocks use renewable raw materials such as palm oil for making biodiesel and including waste oils and fats.

- Hydrazine for use in polymer foams was made using ammonia and sodium hypochlorite but hypochlorite can be replaced by hydrogen peroxide so that no salt co-product is formed and no extra solvent is needed.

- Using liquid waste carbon dioxide under pressure (supercritical) as a residue-free solvent, e.g. in decaffeinating coffee beans, dry cleaning and blowing foam plastics.

- Simvastatin (a cholesterol-reducing drug) is made from cheap raw material using an engineered enzyme made from *E-coli*. Avoids multi-step synthesis using toxic reagents and gives higher yield.

- Poisonous ethylene glycol antifreeze can be replaced by propylene glycol (propane-1,2-diol) formed using glycerol waste from biodiesel production using a copper chromite catalyst that is efficient at lower temperatures (200°) and pressure. It is also used for polyester resins.

- 1,3-propanediol made from corn syrup using genetically modified *E.coli* uses 40% less energy and emitting 20% less greenhouse gas than earlier processes. It is also made from glycerol waste in biodiesel production using bacteria and is used in making polyesters.

- Ionic liquid catalysts convert cellulose into formic acid under oxygen.

- Corn starch is converted to useful resin for plastics by microorganisms.

- Prevent pollution by toxic nitrogen oxides (NO_x) in car/diesel exhausts by adding ammonia to give the redox reaction.

$$4NH_3 + 3NO_2 = 7/2\ N_2 + 6H_2O$$

▲ Biomethane

▲ Palm oil

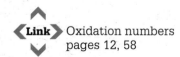

Link Oxidation numbers pages 12, 58

Unit 2

2.4
Organic compounds

Organic compounds are hugely important in everyday life, in understanding the reactions within living species and in industrial processes. To allow new compounds to be synthesised, and their properties linked to potential uses, it is vital that the patterns of behaviour, linked to the structure of the molecules, are recognised. These patterns include the nature of the reactions that the compounds undergo.

International agreement on how compounds are named ensures that scientists throughout the world recognise the compound being described.

You should be able to demonstrate and apply your knowledge and understanding of:

- How to represent simple organic compounds using shortened, displayed and skeletal formulae.

- Nomenclature rules relating to alkanes, alkenes, halogenoalkanes, alcohols and carboxylic acids.

- The effect of increasing chain length and the presence of functional groups on melting/boiling temperature and solubility.

- Isomerism.

- Descriptions of species as electrophiles, nucleophiles and radicals and bond fission as homolytic or heterolytic.

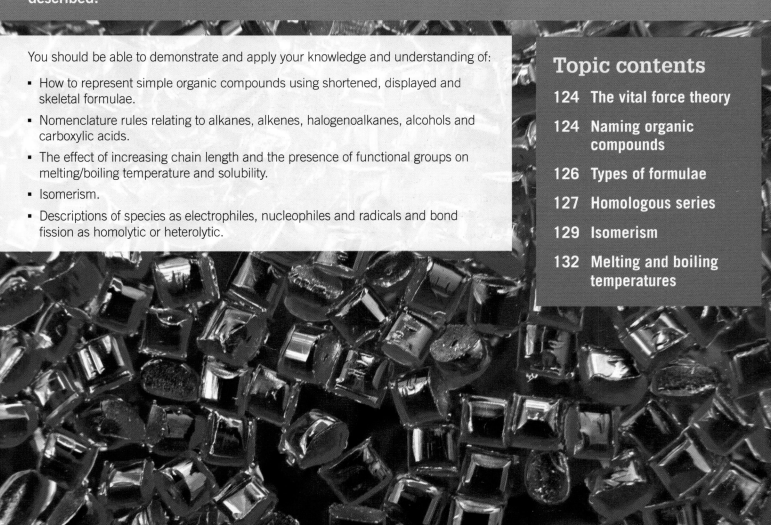

The vital force theory

Up to the 19th century it was thought that chemicals that were present in living organisms could only be formed under the influence of a mysterious force. This force was called the vital force and, because the force could not be created artificially, it was thought impossible to synthesise the chemicals that existed in living organisms in the laboratory. Chemicals produced by living organisms were called 'organic' since this meant 'pertaining to life' whilst other chemicals were called 'inorganic'.

In 1828 a German chemist called Wöhler extracted urea, $(NH_2)_2CO$, from horses' urine and also prepared it in a laboratory from inorganic sources. The term organic therefore lost the 'life' connection and is now used for compounds of carbon in general. There are many thousands of such compounds and whole textbooks have been written about them.

All organic compounds contain carbon and hydrogen and some also contain oxygen, nitrogen, sulfur, phosphorous and the halogens. The elements and their arrangement in a compound define the **functional group** and the functional group defines to which **homologous series** the compound belongs.

Key Terms

Functional group refers to the atom/group of atoms that gives the compound its characteristic properties.

An homologous series is a series of compounds with the same functional group.

Saturated compound is one that contains no C to C multiple bonds.

Hydrocarbon is a compound of carbon and hydrogen only.

Unsaturated compound is one that contains C to C multiple bonds.

▲ Organic or inorganic?

Naming organic compounds

Since there are so many organic compounds it is essential that a system exists that gives each one a name on which all chemists agree. To name a compound you have to know the homologous series to which it belongs. The homologous series in this section are:

Alkanes – **saturated hydrocarbons**

Alkenes – **unsaturated** hydrocarbons with a C to C double bond

Halogenoalkanes – compounds in which one or more hydrogens in an alkane have been replaced by a halogen

Alcohols – compounds containing —OH as the functional group

Carboxylic acids – compounds containing —COOH as the functional group

Stretch & Challenge

Why are there so many compounds of carbon?

Since carbon can form four covalent bonds, it can form chains with itself in different ways. Carbon can also form multiple bonds with itself and with other elements.

You also need to know the 'code' that applies to the number of carbon atoms.

number of carbons	'code'		number of carbons	'code'
1	meth		6	hex
2	eth		7	hept
3	prop		8	oct
4	but		9	non
5	pent		10	dec

Rules for naming compounds

1   Find the longest carbon chain. Using the code above, this is the basis of the name.

2   Number the C atoms in the chain, starting from the end that gives any side chains or functional groups the smallest numbers possible.

3   If there is more than one side chain or substituted group the same, use the prefix di for 2, tri for 3 and tetra for 4.

4   Keep the alphabetical order of branch name.

5   A —CH_3 group is called methyl, a —C_2H_5 group is called ethyl, etc.

6   A saturated hydrocarbons is shown by -ane, a C=C is shown by -ene, an —OH is shown by –ol, a halogen is shown by halogeno and a —COOH is shown by -oic acid.

There is no need to be able to state these rules but you do need to be able to apply them to naming particular compounds.

Examples of how to name compounds

The longest carbon chain has 4 carbons and so the name is based on but.

It is saturated so name ends in -ane.

Numbering from end to give lowest number for side chain gives —CH_3 on carbon number 2.

Name is **2-methyl butane**.

It does not matter which way you number, there are still 3 carbons in the longest chain so the name is based on prop.

There are 2 —CH_3 groups on number 2 carbon and a chlorine on number 1 carbon.

Name is **1-chloro 2,2-dimethyl propane**

18

Knowledge check

Find and number the longest C chain in the compound below

What is the 'code' that applies to the name of this compound?

19

Knowledge check

Draw the formulae of:

(a) 1-chloro but-3-ene

(b) 2,2-dimethyl pentan-3-ol

20

Knowledge check

What are the names of the following?

(a)

(b)

YOU SHOULD KNOW ›››

››› how to show different types of formulae

Key Terms

Molecular formula: shows the atoms, and how many of each type there are, in a molecule of a compound.

Displayed formula: shows all the bonds and atoms in the molecule.

Shortened formula: shows the groups in sufficient detail that the structure is unambiguous.

Skeletal formula: shows the carbon/ hydrogen backbone of the molecule as a series of bonds with any functional groups attached.

There are 5 carbons in longest carbon chain and numbering from the left-hand end gives side chain and functional group the lowest numbers. Name based on pent.

Presence of C═C (alk**ene**) shown by ene (on C 1) and OH (alcoh**ol**) shown by ol (on C 3).

Name is **pent-1-ene,3-ol**

There are 5 carbons in the longest carbon chain – this is if you count along the **bent** chain. Name based on pent.

Presence of —COOH shown by oic acid (this C is C 1) and —CH$_3$ on C 3.

Name is **3-methyl pentanoic acid.**

Types of formulae

The formula of a particular compound can be shown in several ways. The way chosen usually depends on the use being made of the formula.

Molecular formula: shows the atoms, and how many of each type of atom there are in a molecule of compound. All the atoms of the same element are drawn together and therefore the functional group is not always obviously recognisable.

Displayed formula: shows all the atoms, and the bonds linking them, in a compound. This then shows the functional group present clearly and would be used when considering the mechanisms of organic reactions.

Shortened formula: shows the functional group and structure in sufficient detail so that the compound is unambiguous. It cannot be used if details of the bonds are needed.

Skeletal formula: shows the functional groups without the distraction of unreactive chains. It can reduce confusion when complex molecules are being considered and is widely used in research work.

Examples

1 Using 1-chloro 2,2 -dimethyl propane

Molecular formula: $C_5H_{11}Cl$

Displayed formula:

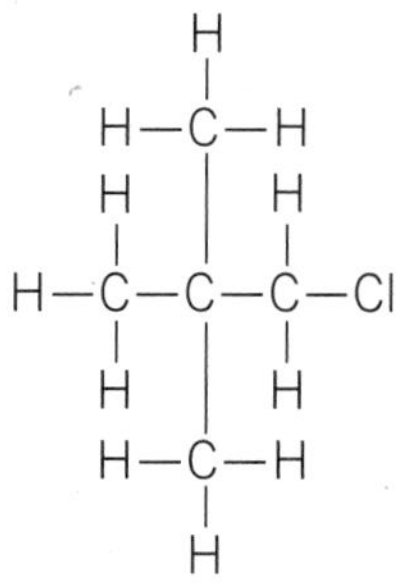

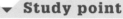

Shortened formula:$CH_3C(CH_3)_2CH_2Cl$

Skeletal formula:

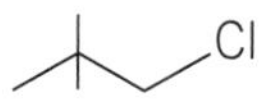

2 Using pent-1-ene, 3-ol

Molecular formula: $C_5H_{10}O$

Displayed formula:

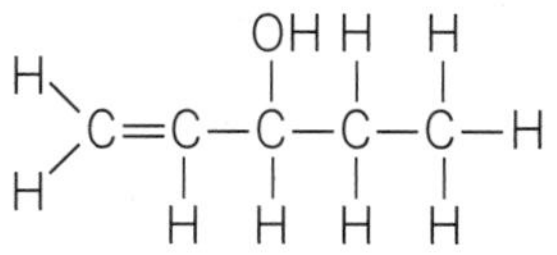

Shortened formula: $CH_2CHCH(OH)CH_2CH_3$

Skeletal formula:

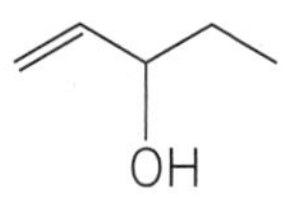

Homologous series

As described earlier, each compound belongs to a particular homologous series. An homologous series is a set of compounds that:

1 Can be represented by a general formula.

2 Differ from their neighbour in the series by CH_2.

3 Have the same functional groups and so very similar chemical properties.

4 Have physical properties that vary as the M_r of the compound varies.

The homologous series in this section are alkanes (saturated hydrocarbons), alkenes (functional group C═C), alcohols (functional group ─OH), halogenoalkanes (functional group ─F, ─Cl, ─Br, ─I) and carboxylic acids (functional group ─COOH).

▼ **Study point**

In an exam, make sure that you use the type of formula that the question asks for – molecular, displayed, shortened or skeletal. There could also be empirical formulae – these are covered later!

21

Knowledge check

The shortened formula of a compound is shown below $CH_3CCl(OH)CHCH_2$

(a) What is its molecular formula?

(b) Draw its displayed formula.

(c) Draw its skeletal formula.

▼ **Study point**

In a skeletal formula, carbon atoms are never shown and only hydrogen atoms within the functional group are shown.

S&C Stretch & Challenge

Why could the skeletal formula of methane be confusing? It's just like a full-stop.

YOU SHOULD KNOW ›››

››› the characteristics of an homologous series

An example of a general formula is shown using the alkanes. Using displayed formulae
the series is:

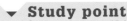

This gives molecular formulae of CH_4, C_2H_6, C_3H_8.

This means that the general formula is C_nH_{2n+2} where n is an integer. For example, if $n = 3$
that means that $2n + 2 = 6$ so formula is C_3H_8.

The effects of the functional group and the way in which physical properties, particularly
meting temperature, boiling temperature and solubility in water, vary within an
homologous series are considered later.

Empirical formulae

The **empirical formula** shows the formula of a compound with the atoms of the elements
in their simplest ratio. This may be the molecular formula but the molecular formula could
also be any multiple of the empirical formula.

Example

Molecular formula of ethane = C_2H_6

Ratio C:H in molecular formula = 2:6

Simplest ratio C:H = 1:3

Empirical formula of ethane = CH_3

Determining an empirical formula

For organic substances most experimental methods are based on burning a known mass
of the compound in excess oxygen and measuring the masses of water and carbon
dioxide produced. This allows the mass of carbon and hydrogen to be calculated but not
any oxygen present – this is considered in the calculation below. You do not need to know
how the halogen content of a compound is determined.

Sometimes questions quote the percentages of elements present. These percentages
can be used as equivalent to mass since they would be the mass of the element in 100g
of compound. The ratio of the atoms of each element is then found using the formula
number of moles = $\dfrac{\text{mass}}{A_r}$.

Example

A compound has the following percentage composition: C = 12.78%; H = 2.15%;
Br = 85.07%. Find its empirical formula.

Ratio C:H:Br = $\dfrac{12.78}{12} : \dfrac{2.15}{1} : \dfrac{85.07}{80}$ = 1.07:2.25:1.06

To find ratio as integers, divide by smallest number – in this case 1.06.

Ratio C:H:Br = 1:2:1.

Empirical formula is CH_2Br.

This calculation **could** give the molecular formula, although not in this case since no such
compound is possible. The molecular formula must therefore be a multiple of this empirical
formula and you need to find which multiple. For this the M_r of the compound is needed.

In this case you are told to assume the M_r is 188.

M_r of empirical formula = 94 so the molecule must be twice the empirical formula.

Molecular formula = $C_2H_4Br_2$

Sometimes, however, a calculation quotes the experimental data and in this case you need to calculate the percentages of the elements from these.

Example

Compound **X** contains carbon, hydrogen and oxygen. 1.55 g of compound X was burned in excess oxygen and 1.86 g of water and 3.41 g of carbon dioxide were formed.

(i) Calculate the empirical formula of **X**.

In CO_2 $\frac{12}{44}$ of the mass is the mass of carbon

3.41 g of CO_2 contains $3.41 \times \frac{12}{44}$ g of C = 0.93 g

% C = $\frac{0.93}{1.55} \times 100$ = 60.0%

In H_2O $\frac{2}{18}$ of the mass is the mass of hydrogen

1.86 g of H_2O contains $1.86 \times \frac{2}{18}$ = 0.207 g

% H = $\frac{0.207}{1.55} \times 100$ = 13.3%

Total % C and H = 73.3%. Rest is oxygen so % O = 27.7%

Ratio C:H:O = $\frac{60.0}{12} : \frac{13.3}{1} : \frac{27.7}{16}$ = 5.0:13.3:1.67

Divide by smallest C:H:O = 3:8:1

Empirical formula = C_3H_8O

(ii) The M_r of **X** is approximately 65. What is the molecular formula of **X**?

M_r of empirical formula = 36 + 8 + 16 = 68.

Molecular formula of **X** = C_3H_8O

Isomerism

There are two types of isomerism in this section – structural and *E-Z*.

Structural isomers are compounds with the same molecular formula but different structural formulae, i.e. arrangement of the atoms.

Structural isomerism can arise in several ways.

1 Chain isomerism

This occurs when the carbon chain of the molecule is <u>arranged differently</u>. Usually one isomer has a straight chain and others have branched chains.

$CH_3CH_2CH_2CH_3$ butane

and

25

Knowledge check

A sample of a hydrocarbon was burned completely in oxygen. 0.660 g of carbon dioxide and 0.225 g of water were formed. The M_r of the compound was approximately 80. Find its empirical formula and hence its molecular formula.

⟨**Link**⟩ The M_r of a compound can be found using mass spectrometry. See page 161

YOU SHOULD KNOW ›››

››› how to recognise structural isomers

Key Term

Structural isomers are compounds with the same molecular formula but with different structural formulae.

▼ Study point

You will be expected to recognise, name and draw isomers of all the homologous series in this section.

Knowledge check 26

Draw displayed formulae for structural isomers of pentane. Name the isomers.

Knowledge check 27

Draw skeletal formulae for two structural isomers of pentene that involve different positions of the double bond. Name the isomers.

Knowledge check 28

Which of the compounds A, B and C are isomers of each other?

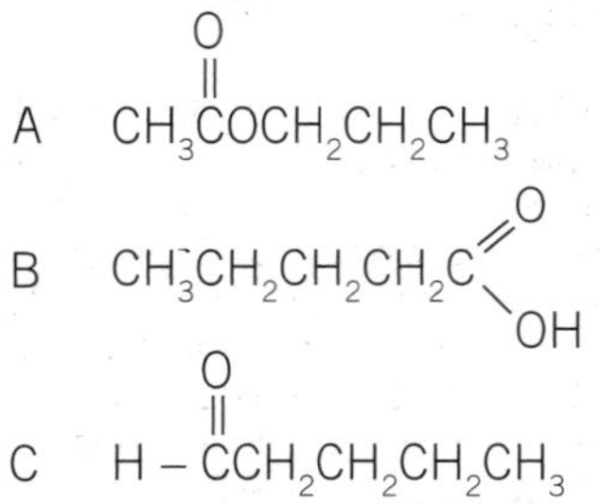

A   CH₃COCH₂CH₂CH₃

B   CH₃CH₂CH₂CH₂C(=O)OH

C   H–COCH₂CH₂CH₂CH₃

YOU SHOULD KNOW ›››

››› why E-Z isomerism occurs

››› how to name E-Z isomers

››› how E-Z isomerism affects properties

Link Link to alkenes, page 139

Key Term

E-Z isomerism is isomerism that occurs in alkenes (and substituted alkenes) due to restricted rotation about the double bond.

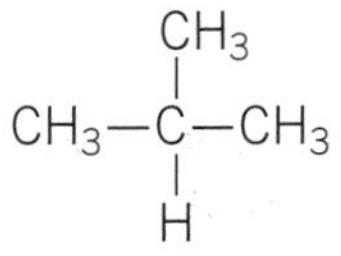

CH₃—C(CH₃)(H)—CH₃

2-methylpropane

2 Position isomerism

This occurs when the functional group is in a different position in the molecule.

CH₂ClCH₂CH₃ 1-chloropropane

and

CH₃CHClCH₃ 2-chloropropane

3 Functional group isomerism

This occurs when the functional group in the compounds is different.

CH₃CH₂OCH₃ an ether

and

CH₃CH₂CH₂OH propan-2-ol

You need to be able to recognise structural isomers but do not need to be able to classify them into chain, position or functional group. Often they involve more than one type.

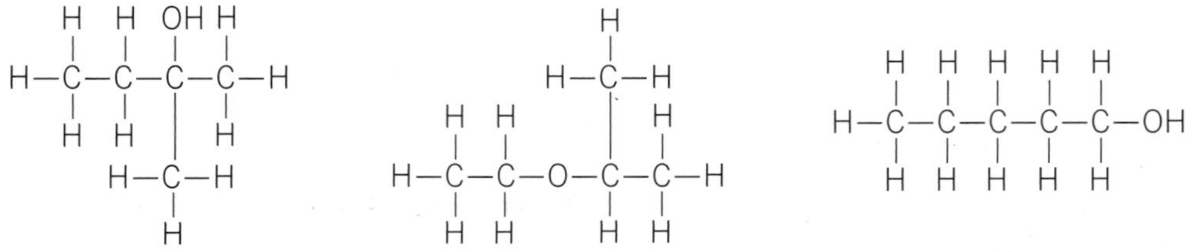

These compounds are all isomers of one another.

E-Z isomerism

Single bonds in alkanes allow the atoms/groups at either end to rotate freely but the double bond in alkenes means that rotation is restricted. This is due to the π bond.

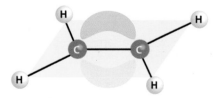

Since the double bond in alkenes and substituted alkenes restricts rotation compounds such as 1,2-dibromoethene can exist in two different forms, i.e. two different isomers. These isomers are called *E-Z* isomers.

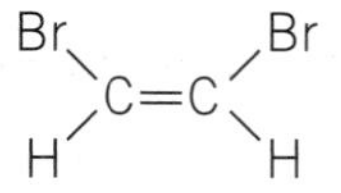

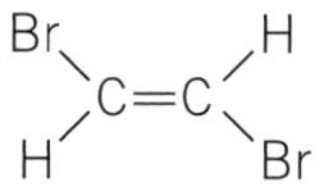

Naming *E-Z* isomers

To decide which isomer is which you must look at the atom directly attached to each carbon in the double bond. Look at each carbon separately and the atom with the higher atomic number takes priority.

In 1,2-dibromoethene C 1 has a bromine and a hydrogen attached. Bromine has a higher atomic number and therefore takes priority.

C 2 also has a bromine and a hydrogen attached. Bromine has a higher atomic number and therefore again takes priority.

If both groups with higher priority are on the same side of the double bond the isomer is *Z*. This means that

$$Br\diagdown C=C\diagup Br$$
$$H\diagup \qquad \diagdown H$$

is (*Z*)-1,2-dibromoethene.

If the groups with higher priority attached to the two carbons are on opposite sides of the double bond this is the *E* isomer.

$$Br\diagdown C=C\diagup H$$
$$H\diagup \qquad \diagdown Br$$

This is (*E*)-1,2-dibromoethene.

The naming system can apply to compounds containing double bonds with different groups attached.

$$Br\diagdown C=C\diagup Cl$$
$$H\diagup \quad_1 \quad _2 \diagdown H$$

Looking at C 1, bromine has a higher atomic number than hydrogen and therefore takes priority. In C 2 chlorine has a higher atomic number than hydrogen and takes priority. Both atoms with higher priority are on the same side of the double bond so that this is (*Z*)-1-bromo 2-chloroethene.

Properties of *E-Z* isomers

Chemical properties

Since the double bond is still present, both isomers will take part in characteristic reactions of alkenes that are dependent on this.

g only H and C

However, the restricted rotation about the double bond means that the substituent group on the double-bonded carbon atoms can behave differently. This is often because in the *E* form the substituent groups are held too far apart to interact but in the *Z* form they are closer.

An example of this is the reaction in which (*Z*)-butenedioc acid can lose water within one molecule.

! Extra Help

You may find a mnemonic for remembering which is which for *E* and *Z*. If you are a linguist, *Z* comes from *zusammen* (= together) and *E* comes from *entgegen* (= opposite). You may prefer *E* for enemies since they are certainly opposite each other.

▼ **Study point**

In older textbooks you will see these types of isomers called *cis-trans*. This has now been superseded by *E-Z* since this has a wider scope of application.

Stretch & Challenge

The naming system can be extended to more complex molecules by looking further along the substituent chains.

Consider

$$CH_3\diagdown C=C\diagup CH_2CH_3$$
$$CH_3CH_2\diagup \quad_1 \quad _2 \diagdown CH_3$$

Looking at C 1, it is attached directly to two carbon atoms – no priority can be assigned. You then have to look at these carbon atoms. In the CH_3 group the C is attached to H H H but in the CH_3CH_2 group this carbon is attached to C H H. This has a higher atomic number and so has priority. This isomer is therefore the *E* form.

! Extra Help

Do not say that there is **no** rotation about the double bond. Rotation is possible but the large input of energy needed means that it is severely restricted.

29

Knowledge check

Draw the *E* isomer of 1-bromo 2-chloroethene.

30

Knowledge check

Does the structure below show the *E* or *Z* isomer?

$$CH_3 \quad Cl$$
$$\diagdown C=C \diagup$$
$$H \diagup \quad \diagdown CH_2CH_3$$

31

Knowledge check

(a) Give the displayed formula of the product of the reaction of bromine with:
   (i) (*Z*)-1,2-dibromoethene
   (ii) (*E*)-1,2-dibromoethene

(b) How are the products of the reactions in (i) and (ii) related to each other?

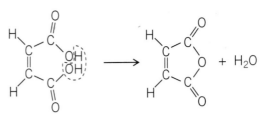

(*Z*)-Butenedioic acid

This reaction is not possible in the *E* form.

(*E*)-Butenedioic acid

Physical properties

Melting temperature and boiling temperature depend on the strength of intermolecular forces. The strength of these is affected by how well the molecules pack together because this alters the area of contact between molecules.

In general the shape of *E* molecules means that they can fit together more closely. This means that they have stronger intermolecular forces and therefore higher melting temperatures. (*Z*)-butenedioc acid has a melting temperature of 130 °C whilst that of (*E*)-butenedioc acid is 286 °C.

Melting and boiling temperatures

The effect of chain length on melting and boiling temperatures

In a solid, the particles are held together rigidly and they can only vibrate about a fixed position. When the solid melts, the forces that hold it rigidly have to be overcome. Although there is less order in a liquid and the particles are further apart, significant attractive forces are still present and these have to be overcome when the liquid changes into a gas. This means that energy is needed to overcome forces whenever a substance melts or boils. This energy is generally in the form of heat. Which substance has a higher melting or boiling temperature can be predicted by looking at the strength of these forces.

Hydrocarbons are simple covalent molecules consisting of only carbon and hydrogen. Since the electronegativities of carbon and hydrogen are similar, hydrocarbons are non-polar. This means that only temporary dipole–temporary dipole **Van der Waals forces** are present between the molecules. These are weak intermolecular forces.

Intermolecular temporary dipole–temporary dipole forces act between the surfaces of the molecules. The more surface there is in contact the more forces will act. This means that more energy is needed to overcome the forces and the melting and boiling temperatures are higher.

YOU SHOULD KNOW ›››

››› how the chain length and amount of branching affects melting and boiling temperatures

 Key Term

Van der Waals forces are dipole–dipole or temporary dipole–temporary dipole interactions between atoms and molecules.

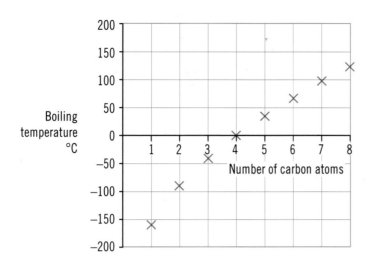

▲ bpt vs chain length

Small hydrocarbons are gases at room temperature, larger ones are liquids and even larger ones are solids.

The effect of branching on boiling temperatures

Above the effect of chain length on boiling temperature was considered so that, in general, the higher the relative molecular mass the higher the boiling temperature. However, different structural isomers will have different boiling temperatures because they have different surface areas that can be in contact.

Consider pentane and its isomer 2,2-dimethylpropane.

Pentane   2,2-Dimethylpropane

Looking at the shapes of these molecules, the more branches that are present the more spherical is the molecule. When spheres are packed together the area available for surface area contact is very small whereas 'sausage-shaped' molecules have much more surface area available.

This means that the more branches an isomer has, the more like a sphere it is and the lower is its boiling temperature.

For pentane and 2,2-dimethyl propane the straight chain pentane has a boiling temperature of 36 °C whilst the branched chain 2,2-dimethylpropane has a boiling temperature of 10 °C.

 **Link** 1.4 Bonding

! Extra Help

Make sure you understand the difference between **inter**molecular and **intra**molecular forces. Intermolecular are between molecules and intramolecular forces are within the same molecule.

32

Knowledge check

Draw **two** molecules of water showing all the bonds. Mark on this diagram an intermolecular force and an intramolecular force.

 Stretch **& Challenge**

Temporary dipole–temporary dipole forces are due to electrons being temporarily on one side of the molecule. The strength of the forces therefore increases as the number of electrons surrounding the molecule increases. This effect, as well as the increased surface area for contact, explains the increase in melting and boiling temperature with increasing chain length.

33

Knowledge check

The boiling temperature of pentane is 36 °C. Suggest a value for the boiling temperature of hexane.

Many branches – little surface contact

Straight chain – more surface area contact

Unit 2

2.5
Hydrocarbons

Hydrocarbons are the naturally occurring raw material for the production of many organic compounds. Petroleum, containing hydrocarbons, can be fractionally distilled to produce fuels suitable for use in a wide variety of heating and other energy-dependent scenarios. It is also the starting point for the production of plastics and other synthetic materials.

In a world where consideration of the environmental issues associated with energy production is increasingly significant, it is necessary to thoroughly understand the nature of the reactions involved.

Topic contents

You should be able to demonstrate and apply your knowledge and understanding of:

- Combustion reactions of alkanes, and benefits and drawbacks relating to the use of fossil fuels, including formation of carbon dioxide, acidic gases and carbon monoxide.
- C—C and C—H bonds in alkanes as σ bonds.
- Mechanism of radical substitution, such as photochlorination of alkanes.
- Difference in reactivity between alkanes and alkenes in terms of the C=C bond as a region of high electron density.
- C=C bond in ethene and other alkenes as comprising π-bond and σ-bond.
- E-Z isomerism in terms of restricted rotation about a carbon to carbon double bond.
- Mechanism of electrophilic addition, such as in the addition of Br_2 to ethene, as a characteristic reaction of alkenes.
- Bromine/bromine water and potassium manganate(VII) tests for alkenes of the possible carbocations involved.
- Nature of addition products of alkenes in terms of the possible carbocations involved.
- Conditions required for the catalytic hydrogenation of ethene and the relevance of this reaction.
- Nature of addition polymerisation and the economic importance of the polymers of alkenes and substituted alkenes.

Fossil fuels

Use of fossil fuels

Much of the world's industrial and domestic energy needs are met by the use of **fossil fuels**. These include natural gas, petroleum and coal. Although progress is being made in developing other sources, it seems we will be dependent, to an appreciable extent, on fossil fuels for the foreseeable future.

Advantages of the use of fossil fuels

1 They are available in a variety of forms so that the type of fuel can be matched with its use. Coal and natural gas, for example, are widely used in power stations. Petroleum can be separated into fractions whose properties can be varied according to the purpose to which the fraction is to be put.

2 They are available at all times. Some green sources such as solar and wind have limited availability.

Disadvantages of the use of fossil fuels

1 They are, in practical terms, **non-renewable**. Theoretically since animals and plants still die, they can be renewed. However, they are considered non-renewable since the formation of fossil fuels takes millions of years and reserves are being used faster that new ones are formed. This means that resources are running out.

2 Combustion of fossil fuels produces carbon dioxide. This is a **greenhouse gas** that acts by absorbing infrared radiation from the earth's surface and then emitting it in all directions. Some of this radiation goes back towards the earth's surface and therefore the surface temperature rises. This climate change has serious environmental consequences including rising sea levels and changes to crop suitability.

▲ Is this caused by global warming?

YOU SHOULD KNOW ›››

››› the nature of fossil fuels

››› the advantages and disadvantages of using fossil fuels

Key Terms

A **fossil fuel** is one that is derived from organisms that lived long ago.

Non-renewable resources are those that cannot be reformed in a reasonable timescale.

A **greenhouse gas** is one that causes an increase in the Earth's temperature.

Stretch & Challenge

Whilst carbon dioxide is usually quoted as the 'villain' in global warming, methane has 34 times the effect on temperature. The largest greenhouse effect is actually due to the presence, in the atmosphere, of water vapour.

! Extra Help

Do not describe ozone depletion when asked about global warming. Ozone depletion is caused by CFCs.

▼ Study point

Many different equations can be used to show the formation of nitric acid in acid rain. These could include:

$N_2(g) + O_2(g) \longrightarrow 2NO(g);$

$2NO(g) + O_2(g) \longrightarrow 2NO_2(g);$

$2NO_2(g) + H_2O(l) \longrightarrow HNO_3(aq) + HNO_2(aq).$

You do not need to be able to quote a specific set of reactions in an exam.

34

Knowledge check

Name two greenhouse gases present in the atmosphere.

35

Knowledge check

Complete the sentence.
The presence of .... acid and
.... acid in rain water ....
the pH.

36

Knowledge check

Complete the equation to show what happens when acid rain runs over a marble statue.

$CaCO_3(s) +$ ___ $HNO_3(aq) \longrightarrow$
.... + ... + ...

Key Term

Acid rain is rain with lower than expected pH.

Study point

If incomplete combustion occurs, water is still always formed. It is the carbon part of the fuel that reacts with less oxygen so that carbon monoxide or even carbon is formed.

37

Knowledge check

Why do cars that are incorrectly adjusted give out black smoke from their exhaust pipes?

38

Knowledge check

Write the equation to show the formation of carbon monoxide from the combustion of butane.

▲ The effects of acid rain

3 Acid rain

Many fossil fuels contain sulfur. On combustion this produces sulfur dioxide. This reacts with water to make sulfuric(IV) (sulfurous) acid.

$$H_2O(l) + SO_2(g) \longrightarrow H_2SO_3(aq)$$

This is then oxidised to make sulfuric (VI) acid.

$$H_2SO_3(aq) + \tfrac{1}{2} O_2(g) \longrightarrow H_2SO_4(aq)$$

At the high temperatures of an internal combustion engine atmospheric nitrogen and oxygen react to form oxides of nitrogen. These react with water to form nitric acid, HNO_3.

Sulfuric and nitric acids are present in **acid rain**. This can cause serious damage to buildings especially those that contain calcium carbonate.

The presence of sulfur dioxide and oxides of nitrogen is a health problem for people with breathing difficulties. During the Beijing Olympics, for example, road traffic was limited to avoid too much air pollution.

4 Carbon monoxide formation

When fossil fuels are burned in an adequate supply of oxygen complete combustion occurs and carbon dioxide and water are formed. However, if there is a shortage of oxygen, incomplete combustion occurs and carbon monoxide is formed. Carbon monoxide is toxic since it combines with haemoglobin in the blood so that this is then not available to carry oxygen around the body.

The need to have adequate oxygen supplies for combustion means that air vents are fitted near boilers and instructions are attached to heaters that they should not be covered.

Incomplete combustion is also less efficient than complete combustion so that, for example, a car would achieve fewer miles per litre of fuel.

Alkanes

The homologous series of alkanes

1 The general formula of the homologous series is C_nH_{2n+2}.
2 Each member of the series differs from its neighbour by CH_2.
3 Alkanes have similar chemical reactions as they are saturated hydrocarbons.
4 The physical properties vary as the relative molecular mass increases. Small alkanes are gases at room temperature (e.g. natural gas is predominantly methane), larger ones are liquids (e.g. petrol contains approximately 8 carbon atoms per molecule) and even larger ones are solids (e.g. wax candles).

Reactions of alkanes

Since the electronegativities of carbon and hydrogen are similar, alkanes are non-polar. They also have no multiple bonds and so are generally unreactive. They do, however, take part in two important reactions.

1 Combustion

Alkanes burn by reaction with oxygen. This reaction is exothermic so that alkanes are used as fuels.

If sufficient oxygen is present **complete combustion** occurs and carbon dioxide and water are formed.

Example using ethane:

$$C_2H_6(g) + 3\tfrac{1}{2} O_2(g) \longrightarrow 2CO_2(g) + 3H_2O(l)$$

If insufficient oxygen is present **incomplete combustion** occurs and carbon monoxide is formed.

Example using propane:

$$C_3H_8(g) + 3\tfrac{1}{2} O_2(g) \longrightarrow 3CO(g) + 4H_2O(l)$$

The carbon monoxide formed in incomplete combustion is toxic since it attaches to the haemoglobin in blood and prevents the haemoglobin carrying oxygen around the body.

Incomplete combustion, as well as producing toxic carbon monoxide, is undesirable as it produces less energy than complete combustion.

Incomplete combustion can also form carbon and this is perhaps seen as black smoke when the engine of a diesel lorry is not properly adjusted.

Key Terms

Complete combustion is combustion that occurs with excess oxygen.

Incomplete combustion is combustion that occurs with insufficient oxygen.

▼ Study point

Since alkanes are non-polar they cannot form hydrogen bonds with water. They are therefore insoluble in water. This can be seen when oil floats on top of water.

Stretch & Challenge

Although you might be asked to write or complete an equation showing incomplete combustion really no wholly valid equation can be written. This is because, whenever carbon monoxide is formed, some carbon dioxide will also be formed. It is not possible to predict the ratio of each.

39

Knowledge check

Write the equation for the complete combustion of octane.

Key Terms

Halogenation is a reaction with any halogen.

Initiation is the reaction that starts the process.

Homolytic bond fission is when a bond is broken and each of the bonded atoms receives one of the bonded electrons.

Radical is a species with an unpaired electron.

Propagation is the reaction by which the process continues/grows.

A chain reaction is one that involves a series of steps and, once started, continues.

Termination is the reaction that ends the process.

A reaction mechanism shows the stages by which a reaction proceeds

A substitution reaction is one in which one atom/group is replaced by another atom/group.

2 Halogenation

Halogenation is the reaction between an organic compound and any halogen (member of Group 17).

Alkanes do not react with halogens in the dark. However, if exposed to uv light (often in the form of sunlight) they will react. The reaction takes place in three stages. The example below uses reaction between methane and chlorine.

Stage 1 Initiation

Initiation starts the reaction. Ultraviolet light has sufficient energy to break the Cl—Cl bond homolytically. **Homolytic bond fission** occurs when each of the bonded atoms receives one of the bond electrons so that **radicals** are formed. Radicals are species that contain an unpaired electron. This means that they react indiscriminately to gain another electron and form a pair. In a radical only the unpaired electron is generally shown. Be careful with the terms *unpaired electron* and *lone pair* of electrons. Make sure you know what each means.

$$Cl_2 \longrightarrow 2Cl^\bullet$$

Stage 2 Propagation

Radicals are very reactive and take part in a series of **propagation** reactions.

$$Cl^\bullet + CH_4 \longrightarrow CH_3^\bullet + HCl$$

$$CH_3^\bullet + Cl_2 \longrightarrow CH_3Cl + Cl^\bullet$$

A propagation reaction uses a radical as a reactant and then forms a radical as a product. This means that the reaction continues. It is therefore called a **chain reaction**.

Stage 3 Termination

The propagation steps continue until two radicals meet. The reaction then stops – this is the **termination** stage.

$$Cl^\bullet + CH_3^\bullet \longrightarrow CH_3Cl$$

The stages initiation, propagation and termination are the **mechanism** of the **reaction**.

Exam questions may ask for the mechanism but be careful since sometimes a question asks for the **overall equation**. That is not the same as the mechanism. If, for example, a question asked for the equation for the reaction of ethane with bromine, the expected answer is:

$$C_2H_6 + Br_2 \longrightarrow C_2H_5Br + HBr$$

Since mixtures are always formed, halogenation of alkanes is not a very satisfactory method of preparation. However, polysubstitution can be largely avoided if the amount of halogen used is very limited.

40 Knowledge check

Complete the following:

Alkanes are generally unreactive because .... Radicals are very reactive

because ....

41 Knowledge check

Write the equation for the formation of tetrachlormethane from methane.

42 Knowledge check

Why would you observe 'steamy fumes' when chlorine reacts with an alkane?

The Alkenes

Structure of alkenes

Alkenes are an homologous series of unsaturated hydrocarbons. This means that they contain a carbon to carbon double bond and have the general formula C_nH_{2n}.

Alkenes are formed when petroleum is cracked and, although they will undergo combustion reactions, they are not generally used as fuels. They are, however, very important as the starting material for a variety of organic synthesis reactions including polymerisation. → 2°° mono to form poly

Due to the presence of the double bond, alkenes are much more reactive than alkanes.

Ethene

In ethene each carbon has the electronic structure $1s^2 2s^2 2p^2$, i.e. each has four electrons available for bonding. Three 'normal' covalent bonds are forms using three of these electrons. These bonds are called σ bonds and they involve both s electrons and one of the p electrons in each carbon. This means that there are three areas of negative charge around each carbon and VSEPR theory says that the molecule is planar triangular with a bond angle of 120°. One p electron orbital on each carbon is not used to form these bonds,

trigonal planar

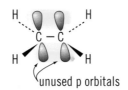

unused p orbitals

The p orbitals overlap **sideways** to produce a π **bond** – an area of high electron density above and below the plane of the molecule.

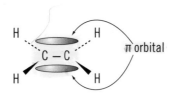

π orbital

Other alkenes

Other alkenes also have bond angles of approximately 120° around the carbon atoms attached to the double bond but bond angles of approximately 109.5° about the carbon atoms in the saturated part of the chain.

Example propene

120°

109.5°

Key Term

A π **bond** is one formed by the sideways overlap of p electrons.

43

Knowledge check

Draw the displayed formula of (Z) pent-2-ene.

44

Knowledge check

Why are alkenes not generally used as fuels?

Stretch & Challenge

Bond formation involving s to s orbital overlap (or any other overlap of orbitals along the axis) produces σ bonds. **Sideways** overlap of p orbitals, however, produces π bonds.

The restricted rotation about double bonds, responsible for E-Z isomerism, is due to the presence of the π bond. Too much energy is needed to break the bond to allow free rotation.

Reactions of alkenes

Electrophilic addition reactions

Alkenes have a pair of electrons in a π orbital that makes them susceptible to attack by an electrophile. An **electrophile** is any species that can accept a lone pair of electrons. The mechanism of the reaction involves **heterolytic bond fission** leading overall to an **addition reaction**.

If, for example, hydrogen bromide is used, this has a permanent dipole and so attacks the π bond with the $\delta+$ H being nearer to the negative area.

The mechanism in this case involves the use of 'curly arrows'. A curly arrow shows the movement of a **pair** of electrons.

If a non-polar molecule, such as hydrogen or bromine, reacts with ethene a dipole is induced by the negative charge in the π bond.

X_2 could be H_2 or Br_2 to produce the overall equations

$$C_2H_4 + H_2 \longrightarrow C_2H_6$$
ethene · · · · · · · ethane

$$C_2H_4 + Br_2 \longrightarrow CH_2BrCH_2Br$$
ethene · · · · · · · 1,2-dibromoethane

Uses of these reactions

(i) Bromine

The reaction with bromine is used as the test for an alkene. Bromine is brown and therefore the presence of an alkene is shown by the decoloration of brown bromine water. Sometimes bromine water is used instead of liquid bromine in this test. This is because liquid bromine is very corrosive.

(ii) Hydrogen

The reaction with hydrogen can be called hydrogenation. It is catalysed by transition metals including platinum and palladium but nickel is most commonly used. The reaction is commercially important because liquid vegetable oils contain many double bonds (they are polyunsaturated) and these can be made saturated by adding hydrogen. This hardens the oil to make solid edible fats (margarines/butter substitutes).

Reaction of propene with hydrogen bromide

Consider the reaction between propene and hydrogen bromide. Two products are possible.

Propene + HBr → 2-Bromopropane Reaction 1

Propene + HBr → 1-Bromopropane Reaction 2

In practice far more of the 2-brompropane is formed than the 1-bromopropane. This can be explained by looking at the mechanism of the reaction.

Study point

In a test involving a change in colour you should include the original colour as well as what is seen in the test.

The decoloration of purple acidified potassium manganate(VII) can also be used as the test for an alkene.

Knowledge check

Suggest two reasons why bromine water, rather than chlorine, is used as the test for an alkene.

Stretch & Challenge

Decoloration of brown bromine is really the test for unsaturation rather than just alkenes. Alkynes, that contain a carbon to carbon triple bond, will also give a positive result.

Study point

Delocalisation (the spread of a charge across several atoms) stabilises the ion. You will see this again when you consider aromatic compounds.

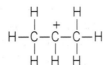

Reaction 1   2° carbocation

Reaction 2   1° carbocation

Key Term

A **carbocation** is a carbon-containing positively charged ion.

46

Knowledge check

Why, in the manufacture of 'butter substitutes', is nickel generally used as a catalyst (rather than platinum or palladium)?

47

Knowledge check

Complete the equation:

$(CH_3)_2C{=}CH_2 + HBr \longrightarrow$

YOU SHOULD KNOW ›››

››› the nature of addition polymerisation

››› how to draw and name addition polymers

››› how to recognise a repeat unit

Key Terms

Polymerisation is the joining of a very large number of monomer molecules to make a large polymer molecule.

Monomer is a small molecule that can be made into a polymer.

▼ **Study point**

The empirical formula of an addition polymer is the same as that of the monomer.

The positive ion (the **carbocation**) is more stable in reaction 1 than in reaction 2. This is because alkyl groups tend to release electrons so that they become δ+. This spreads or delocalises the positive charge and this stabilises the ion. This means that 2° carbocations are more stable than 1° carbocations.

Polymerisation

Polymerisation is the joining of a very large number of **monomer** molecules to make a large polymer molecule. Alkenes, and substituted alkenes, undergo addition polymerisation. This means that the double bond is used to join the monomers and only one product forms.

Alkenes, and substituted alkenes, undergo addition polymerisation. In this type of reaction the double bond is used to join the molecules. When ethene is polymerised poly(ethene) is formed – this is commonly called polythene. In the equation below, showing this polymerisation, *n* is used to show a very large number.

The polymer does not contain a double bond **but** the name is formed by adding poly to the name of the monomer so that it involves ene.

Poly(ethene) was discovered by accident when ethene and small amounts of other compounds were subjected to very high pressures and a white waxy material was formed. In the early days it was difficult to control the polymerisation to make straight chains and the polymer produced had side chains. This meant that the chains could not pack closely together and the density of the poly(ethene) was low. The chains also had few points at which they came close and this meant that Van der Waals forces were limited and therefore the melting temperature was low.

Since poly(ethene) contains only single bonds, it is really a very long chain alkane and this means that it is an unreactive, flexible solid that can be used to make plastic bags, etc. The low melting temperature of branched chain polymers could limit their uses until it was discovered that the use of catalysts (including Ziegler catalysts) could allow the production of straight chain poly(ethene). This meant that the chains could pack together better to give a higher density plastic which was therefore harder. The melting temperature was also much higher so that other uses, where a high temperature or the need for rigidity was involved, were developed.

The differences between straight and branched chains when ethene was polymerised were an early example of how changing the structure of the polymer can change its physical properties to make it suitable for a particular use. This can also be achieved by using substituted alkenes as the monomers.

Polymers of substituted alkenes

Since the monomers are joined using the double bond, polymers of substituted alkenes can be produced.

One method of finding the structure of the polymer formed from a particular monomer is to draw the monomer several times with the double bond in one molecule **next to** that in another molecule. The double bond breaks and is used to join the monomers together. This is shown by the arrows in the diagram.

This gives the polymer as

The **repeat unit** of this polymer is

Economically important polymers include:

(i) Poly(propene)

Monomer propene   Polymer, poly(propene)

Poly(propene) is rigid and used in food containers and general kitchen equipment.

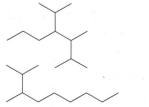
Key Term

Repeat unit is the section of the polymer that is repeated to make the whole structure.

(ii) Poly(chloroethene)

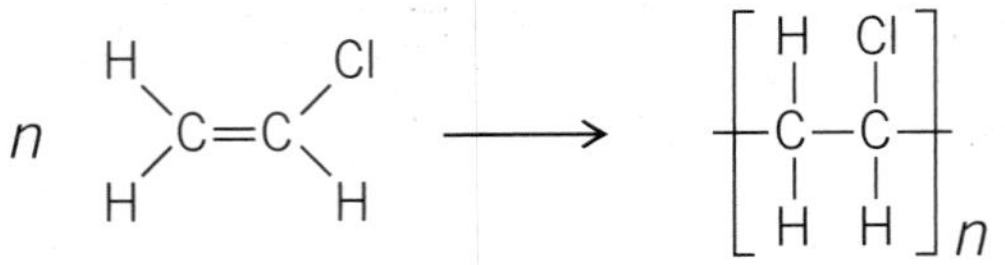

Monomer
chloroethene

Polymer
poly(chloroethene)

48

Knowledge check

What is the name of the polymer formed from 1-bromo, 2-nitroethene?

The old name for chloroethene was vinyl chloride. This means that poly(chloroethene) was called polyvinyl chloride (often abbreviated to PVC).

Poly(chloroethene) can be used in water pipes, in waterproof clothing or as the insulating covering for electrical cable.

49

Knowledge check

Draw 3 repeat units for the polymer you have named in 1. (The nitro group is NO_2.)

50

Knowledge check

What is the empirical formula of poly(ethene)?

(iii) Poly(phenylethene)

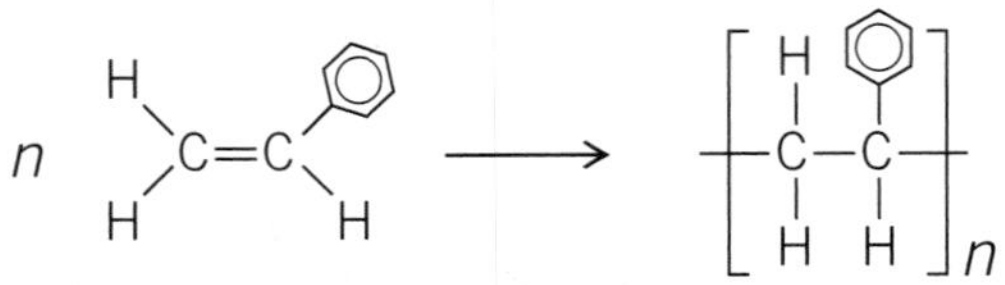

Monomer
phenylethene

Polymer
poly(phenylethene)

The old name for phenylethene was styrene. This means that poly(phenylethene) was called polystyrene. This polymer is hard and is used in many household items needing strength and rigidity. It can be made into an insulator and packing material by creating holes in the structure. This was then called expanded polystyrene.

! Extra Help

When drawing a section of a polymer you must show the — at both ends to show that the chain continues. If you only show one repeat unit, do not forget to put '*n*' outside the bracket.

51

Knowledge check

A section of polymer is shown below. Draw the monomer that was used to form this polymer.

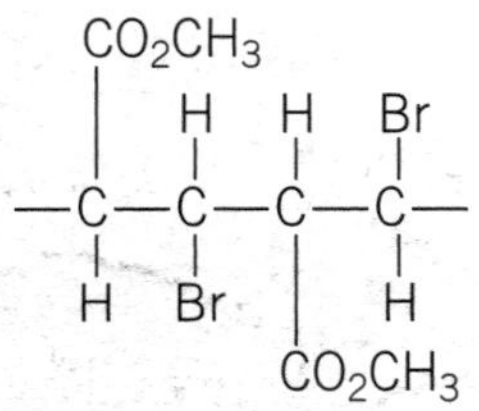

Unit 2

2.6
Halogenoalkanes

Halogenoalkanes are still used for a variety of industrial and household purposes but their use is now strictly controlled due to their toxicity and the environmental damage that they can cause. Understanding the chemistry of halogenoalkanes means that safer alternatives are now being developed.

As an homologous series halogenoalkanes show reactions caused by nucleophilic substitution. These can be used for studies of the effect of changing the nature of the halogen or the structural isomer used on the rates of reactions.

You should be able to demonstrate and apply your knowledge and understanding of:

- Elimination reaction of halogenoalkanes forming alkenes, for example HBr eliminated from 1-bromopropane to form propene.
- Mechanism of nucleophilic substitution, such as the reaction between $OH^-(aq)$ and primary halogenoalkanes.
- Effect of bond polarity and bond enthalpy on the ease of substitution of halogenoalkanes.
- Hydrolysis/$Ag^+(aq)$ test for halogenoalkanes.
- Halogenoalkanes as solvents, anaesthetics and refrigerants, and tight regulation of their use due to toxicity or adverse environmental effects.
- Adverse environmental effects of CFCs and the relevance of the relative bond strengths of C—H, C—F and C—Cl in determining their impact in the upper atmosphere.
- How to carry out a reflux (for example, for nucleophilic substitution reaction of halogenoalkanes with hydroxide ions).

Structure

Halogenoalkanes are an homologous series in which one or more of the hydrogen atoms have been replaced by a halogen. The series has the general formula $C_nH_{2n+1}X$ (where X is a halogen). The formula of a halogenoalkane is often shown as RX. Polysubstituted halogenoalkanes can also exist.

Halogenoalkanes contain a carbon to halogen bond. Since halogens are more electronegative than carbon, this bond is polar with a $\delta+$ carbon and a $\delta-$ halogen.

This dipole means that halogenoalkanes are susceptible to **nucleophilic** attack on the $\delta+$ carbon. This leads to substitution.

Nucleophilic substitution

Mechanism

A nucleophile has a lone pair that can be donated to an electron-deficient centre.

Using OH^- and 1-chloropropane as an example:

1-Chloropropane   Propan-1-ol

Preparation of alcohols

This reaction can be used to prepare alcohols from halogenoalkanes, e.g. butan-1-ol from 1-bromobutane. The nucleophile OH^- is provided by aqueous sodium hydroxide.

$$CH_3CH_2CH_2CH_2Br + OH^- \longrightarrow CH_3CH_2CH_2CH_2OH + Br^-$$

Since the reaction is rather slow it is necessary to heat the mixture of 1-bromobutane and aqueous sodium hydroxide for a significant time. If this were done in an open flask or beaker much of the liquids would evaporate and be lost and the yield of product would be very low. To avoid this problem the liquid mixture is **refluxed**.

The reaction between aqueous hydroxide ions and a halogenoalkane is classified as being nucleophilic substitution. However, since the change could be achieved using the OH^- in water, it can also be classified as **hydrolysis**.

YOU SHOULD KNOW ›››

››› the mechanism of the substitution reaction of halogenoalkanes

››› the hydrolysis reaction of halogenoalkanes

››› the elimination reaction of halogenoalkanes to give alkenes

››› the uses of halogenoalkanes

››› the adverse environmental effects of CFCs

Key Terms

A halogenoalkane is an alkane in which one or more hydrogen atoms have been replaced by a halogen.

A nucleophile is a species with a lone pair of electrons that can be donated to an electron deficient species.

Reflux is a process of continuous evaporation and condensation.

Hydrolysis is a reaction with water to produce a new product.

▼ Study point

There are only really three mechanisms in this section of work. One of them is almost certainly to be on the exam paper – make sure you understand them!

! Extra Help

When liquids are refluxed the top of the condenser must be open. If it is not, when the apparatus is heated, the air will expand and blow the stopper off – usually violently!

lam .

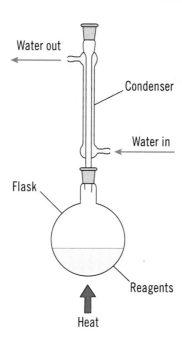

Water out

Condenser

Water in

Flask

Reagents

Heat

When heated the liquids evaporate and vapour escapes from the flask. However when the vapour reaches the condenser it condenses and liquid forms. This liquid drips back to the reaction flask. This means that in this process of continuous evaporation and condensing the liquid can be boiled for as long as is needed to achieve a reaction without any loss of material.

Effect of changing the halogen

A similar reaction occurs for other halogens but the rate at which the reaction occurs depends on the nature of the halogen. Using chloro, bromo and iodo compounds there are two factors that have to be considered.

(i) Electronegativity
Electronegativity decreases as the size of the halogen increases. This means that the C—Cl is the most polar with this carbon atom being the most δ+.

(ii) Bond strength
The substitution reaction involves breaking the carbon to halogen bond. This is strongest for the C—Cl bond and this therefore is the most difficult to break.

The effects in (i) and (ii) act in opposite directions and so it is difficult to predict which of the halogenoalkanes would be hydrolysed most rapidly. In fact (ii) dominates so that the order of rate of hydrolysis for corresponding halogenoalkanes is:

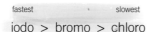

fastest   slowest

iodo > bromo > chloro

The rate of the reaction can be followed using the fact that after hydrolysis a halide ion is formed in solution. This can be detected by adding $Ag^+(aq)$, usually as aqueous silver nitrate, and timing how long the precipitate takes to form.

52 Knowledge check

Name the reagents that could be used to prepare 2-methylbutan-2-ol.

53 Knowledge check

What conditions are needed for this reaction?

54 Knowledge check

Write the equation for the reaction in KC52.

55 Knowledge check

Why is chlorine more electronegative than bromine?

56 Knowledge check

Why is the C—Cl bond stronger than the C—Br bond?

▼ **Study point**

The test for a halogenoalkane is in two parts. First is the formation of the halide ion and this is then followed by testing for this ion.

! **Extra Help**

Even though the question asks about the reaction between a halogeno compound and aqueous silver nitrate, the first stage is the reaction with OH⁻.

57

Knowledge check

Write down, in order, the steps needed to test for the presence of bromine in an organic compound. Include the result expected.

Key Term

An **elimination reaction** is one that involves the loss of a small molecule to produce a double bond.

58

Knowledge check

Write the ionic equation for the reaction that occurs to produce the yellow precipitate formed if iodine is present in an organic compound.

59

Knowledge check

What is the product of the reaction of 1-bromo 2-methylbutane with:

(a) Aqueous sodium hydroxide?

(b) Ethanolic sodium hydroxide?

60

Knowledge check

What is the product of the reaction of 2-bromo 2-methylbutane with:

(a) Aqueous sodium hydroxide?

(b) Ethanolic sodium hydroxide?

Test for halogenoalkanes

Water can be used to hydrolyse the halogenoalkane but this is rather slow and therefore aqueous sodium hydroxide is often used. Before aqueous silver nitrate is added the excess sodium hydroxide must be neutralised by adding dilute nitric acid, as sodium hydroxide would interfere with the test.

$$RX + NaOH(aq) \longrightarrow ROH + Na^+(aq) + X^-(aq)$$

The presence of the halide ion is then shown by the aqueous silver nitrate in the usual test.

$$X^-(aq) + Ag^+(aq) \longrightarrow AgX(s)$$

halogen in halogenoalkane	addition of $Ag^+(aq)$	addition of $NH_3(aq)$ to precipitate formed with $Ag^+(aq)$
chlorine	white precipitate	dissolves in dilute $NH_3(aq)$
bromine	cream precipitate	dissolves in concentrated $NH_3(aq)$
iodine	yellow precipitate	does not dissolve in $NH_3(aq)$

Elimination reactions

An **elimination reaction** is one that involves the loss of a small molecule to produce a multiple bond.

As well as the substitution reactions seen above, halogenoalkanes can undergo elimination reactions. Hydrogen halides are acidic and therefore can be removed using an alkali. The alkali, such as sodium hydroxide, must be dissolved in ethanol to avoid the substitution reaction. Using 1-bromopropane as an example of an elimination reaction is shown in the equation:

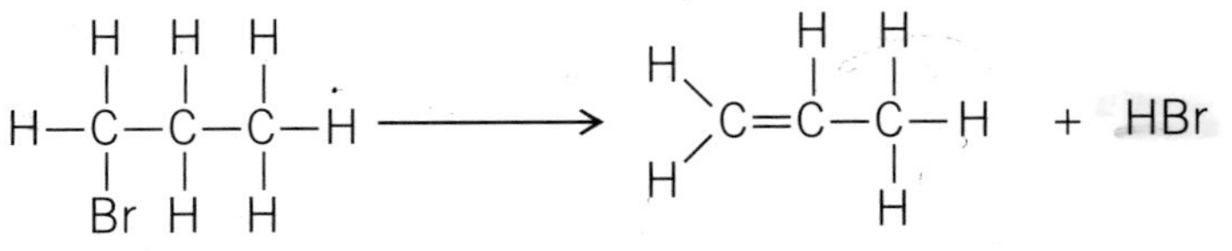

1-Bromopropane   Propene   + HBr

This equation can also be written as:

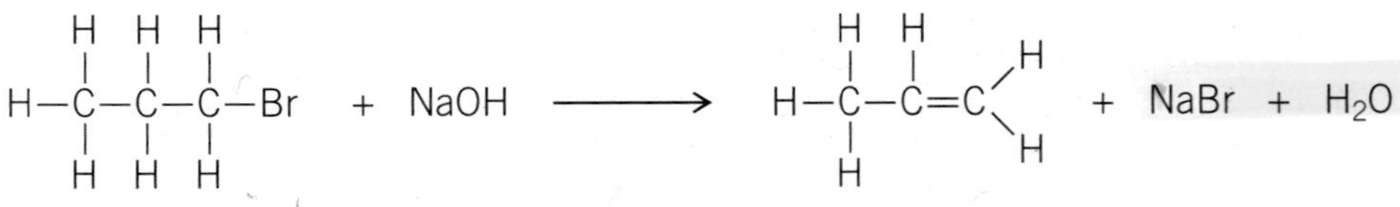

In order to eliminate hydrogen halide, the halogen must be attached to a carbon next to a carbon that has a hydrogen attached.

Example

If the halogenoalkane is unsymmetrical, more than one alkene can be formed by elimination reactions. An example is:

But-1-ene

But-2-ene

You do not need to know which product would dominate.

Uses of halogenoalkanes

1 As solvents

Halogenoalkanes contain a polar section due to the presence of the carbon halogen bond but they also contain a non-polar section due to the presence of the alkyl chain. This means that they can mix with a variety of polar and non-polar organic substances and are therefore used as solvents. The non-flammability of halogenoalkanes is also an advantage so that they can be used in dry cleaning clothes and other processes involving degreasing. Chlorocompounds such as tetrachloromethane, CCl_4, and tetrachloroethene, $CCl_2 = CCl_2$, were widely used.

Chlorocompounds were preferred to other halogenoalkanes because they were cheaper.

2 As anaesthetics

Many halogenoalkanes can act as general anaesthetics. Trichloromethane, $CHCl_3$, (chloroform) was one of the earliest substances to be used and this revolutionised surgical procedures. Another famous 'knock out' drug used in the past was a Mickey Finn. This consists of 2,2,2-trichloroethane-1,1-diol, $CCl_3CH(OH)_2$, (chloral hydrate). Although these compounds are no longer in use halothane, $CF_3CHBrCl$, is still used in anaesthesia.

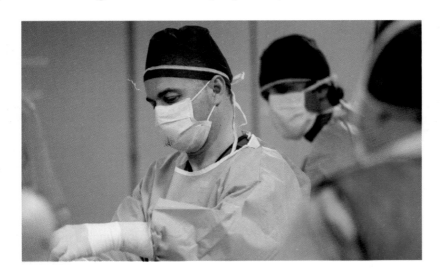

Key Terms

CFCs halogenoalkanes containing both chlorine and fluorine.

Ozone layer a layer surrounding the earth that contains O_3 molecules.

HFCs halogenoalkanes containing fluorine as the only halogen.

Stretch & Challenge

Halogenoalkanes are virtually insoluble in water because they do not contain the — OH or — NH needed to hydrogen bond with the water. They are, however, able to mix with a large variety of organic substances since the change in the strength of the intermolecular forces on mixing is small.

3 As refrigerants

Although small halogenoalkanes are gases at room temperature, the presence of permanent dipole–permanent dipole attractions means that their boiling temperatures are close to room temperature. They are therefore liquids that can easily be evaporated or gases that can easily be liquefied at room temperature. **CFCs** were used as refrigerants since the heat needed to change the liquid to the gas is removed from the fridge to cool its contents. Non-flammability and non-toxicity, as well as a suitable boiling temperature, are important in making halogenoalkanes suitable for this use.

Regulation of uses of halogenoalkanes.

Nowadays the use of halogenoalkanes is limited by statutory regulations. This is because many of them, particularly polychloroalkanes, have been found to be toxic and others, particularly the chlorofluorocarbons, CFCs, cause damage to the earth's **ozone layer**. The use of CFCs is thought to have caused holes in the ozone layer, particularly over the Arctic and Antarctic.

Damage to the ozone layer allows uv rays to reach the earth's surface and cause skin cancer. The damage process involves radical chain reactions and, as seen with halogenations of alkanes, the mechanism includes a series of steps.

▼ **Study point**

Do not try to remember particular propagation stages. Many alternatives are possible but they all convert O_3 to O_2 whilst regenerating the chlorine radical.

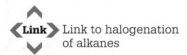 Link to halogenation of alkanes

Initiation stage

In the upper atmosphere uv radiation causes homolytic bond fission of the C—Cl bond in the CFC. Using dichlorodifluoromethane, CCl_2F_2, as an example of a CFC:

$$CCl_2F_2 \longrightarrow Cl^\bullet + CCl^\bullet F_2$$

Propagation stages

There are many possible propagation stages which include:

$$Cl^\bullet + O_3 \longrightarrow ClO^\bullet + O_2$$
$$ClO^\bullet + O_3 \longrightarrow Cl^\bullet + 2O_2$$

Since these form a chain reaction the presence of a small number of chlorine radicals can cause the decrease of many ozone molecules.

Alternatives to CFCs

To avoid the formation of chlorine radicals alternative compounds that do not contain chlorine are being increasingly used. These include alkanes (although these have the problem of inflammability) and **HFCs**. Since the only halogen in HFCs is fluorine, radicals are not formed when they are exposed to uv radiation.

61

Knowledge check

Some bond strengths are given in the table.

Bond	Bond strength/ kJ mol^{-1}
C—H	413
C—Cl	328
C—F	485

Use these data to explain why it is the chlorine radical that is formed when the compound is exposed to uv light.

62

Knowledge check

Why is the C—F bond stronger than the C—Cl bond?

2.7
Alcohols and carboxylic acids

Everyone will be aware of some alcohols and carboxylic acids. To many people 'alcohol' refers to the ethanol that is present in alcoholic drinks but to a chemist the word refers to compounds with the OH functional group.

Carboxylic acids also contain OH but in this case it is within the COOH functional group and this gives the acids very different properties from the alcohols. Carboxylic acids are widely distributed in 'acidic' foodstuffs including vinegar, lemons and sour milk.

The reaction between an alcohol and a carboxylic acid gives an ester – these are widely used as the active part of perfumes and flavourings.

You should be able to demonstrate and apply your knowledge and understanding of:

- Industrial preparation of ethanol from ethene.
- Preparation of ethanol and other alcohols by fermentation followed by distillation, and issues relating to the use of biofuels.
- Dehydration reactions of alcohols.
- Classification of alcohols as primary, secondary and tertiary.
- Oxidation of primary alcohols to aldehydes/carboxylic acids and secondary alcohols to ketones.
- Dichromate(VI) test for primary/secondary alcohols and sodium hydrogencarbonate test for carboxylic acids.
- Reactions of carboxylic acids with bases, carbonates and hydrogencarbonates forming salts.
- Esterification reaction that occurs when a carboxylic acid reacts with an alcohol.
- Separation by distillation.

Alcohols

Alcohols are an homologous series in which one of the hydrogen atoms in the alkane has been replaced by —OH. This means that —OH is the functional group. Alcohols also exist that contain more than one —OH group but ethanol is the most widely used alcohol. This is what, in everyday language, is called alcohol and it is present in alcoholic drinks. The alkyl group can be shown as R so that an alcohol would be ROH.

YOU SHOULD KNOW ›››

››› the preparation of ethanol from ethene

››› the preparation of ethanol by fermentation

››› some issues relating to the use of biofuels

››› the classification system for alcohols

››› the reactions of alcohols

Key Terms

Alcohol an homologous series containing —OH as the functional group.

Fermentation an enzyme-catalysed reaction that converts sugars to ethanol.

Stretch & Challenge

The mechanism for the hydration of ethene involves electrophilic attack on the π bond by the δ+ H on the water. Try and draw this mechanism using curly arrows.

Industrial preparation of ethanol

Since ethanol is industrially and commercially important it is necessary to manufacture it on a large scale. This is achieved in two main processes.

(a) Hydration of ethene
Direct

Ethene is obtained, on a large scale, by cracking hydrocarbons produced from petroleum. It reacts with steam to produce ethanol.

$$CH_2\!\!=\!\!CH_2(g) + H_2O(g) \rightleftharpoons CH_3CH_2OH(g) \qquad \Delta H = -45\,kJ\,mol^{-1}$$

The conditions generally used in this conversion are a temperature of about 300°C, a pressure of about 60–70 atmospheres and a catalyst of phosphoric acid (coated onto an inert solid). These conditions can be explained using Le Chatelier's principle.

Temperature

Since the forward reaction is exothermic a high yield would be favoured by a low temperature. This, however, gives a slow rate of reaction: 300°C is a compromise temperature.

Pressure

In the reaction two moles of gas react to produce one mole of gas. A high yield is therefore favoured by a high pressure. High pressure also increases the rate of reaction but if the pressure is too high more powerful pumps and stronger pipes are needed and this increases costs. 60–70 atmospheres is therefore used,

Catalyst

The catalyst does not affect the yield but it does increase the rate at which ethene and steam react to produce the equilibrium concentration of ethanol.

Under these conditions only about 5% of the ethene is converted and therefore the remaining ethene is recycled back to the reaction chamber.

(b) Fermentation

Fermentation is the process by which sugars are converted into ethanol. The reaction is generally carried out by dissolving the sugar in water, adding yeast and leaving the mixture

in a warm place. Yeast contains enzymes that catalyse the reaction. Using glucose as an example the equation for the reaction is:

$$C_6H_{12}O_6 \longrightarrow 2C_2H_5OH + 2CO_2$$

Fermentation is the method by which alcoholic drinks are produced.

Carbon dioxide escapes as a gas but the ethanol has to be separated from the remaining liquid mixture. The boiling temperature of ethanol is 80 °C and therefore to separate it from an aqueous mixture fractional distillation in needed.

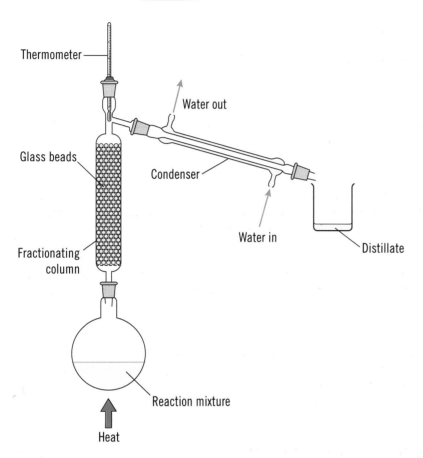

63

Knowledge check

How could ethanol be removed from the equilibrium mixture produced when ethene reacts with steam?

 Extra Help

Nothing, apart from glucose is needed, as a reagent in fermentation. Water is needed for the reaction to proceed but it is not part of the overall equation.

64

Knowledge check

Why is fractional distillation, rather than merely distillation, needed in this separation?

Biofuels

Biofuels are fuels that are produced from living organisms. The two main types currently is use are bioethanol and biodiesel. Bioethanol is obtained from sugars in plants by fermentation (as described above) and biodiesel is obtained from the oils and fats present in the seeds of some plants.

 Key Term

Biofuel a fuel that has been produced using a biological source.

Since both bioethanol and biodiesel burn exothermically they can be used as fuels. If used in internal combustion engines they are generally blended with conventional fossil fuels.

Issues relating to the use of biofuels

There are advantages and disadvantages in using biofuels compared with using fossil-based fuels.

Advantages

(i) Renewable

Fossil fuels are non-renewable and will eventually run out. Plants can be grown each year and biofuels can also be produced using waste material from animals.

(ii) Greenhouse gases

Overall production of carbon dioxide is reduced. Although carbon dioxide is produced in exactly the same way by biofuels and fossil fuels when combusted, carbon dioxide is taken in by the plants when they are grown. Plants use carbon dioxide in photosynthesis to produce sugars.

$$6CO_2 + 6H_2O \longrightarrow C_6H_{12}O_6 + 6O_2$$

This means that, since carbon dioxide is produced in one process but removed in another, overall the use of biofuels can be considered to be carbon neutral.

(iii) Economic and political security

Countries that do not have fossil fuels as a natural resource are less dependent on changes in price and availability that having to import fossil fuels imposes.

Disadvantages

(i) Land use

Land that is used to produce plants for biofuels cannot be used to produce food. There is also pressure to destroy environmentally significant areas, such as forests, to create land for biofuel production.

(ii) Use of resources

Growing crops suitable for biofuels needs large quantities of water and fertilisers. The use of water can strain local resources and the use of large quantities of fertilisers to grow the same crop year after year can cause water pollution.

(iii) Carbon neutrality?

When the fuel needed to build and run the factories needed in biofuel production, to transport raw materials and finished products, etc. are considered, it can be argued that the use of such fuels is not carbon neutral.

Dehydration of alcohols

Many alcohols can be dehydrated to form alkenes. The equation for propan-1-ol is shown below:

$$CH_3CH_2CH_2OH \longrightarrow CH_3CH{=}CH_2 + H_2O$$

propan-1-ol $\qquad$ propene

Many different dehydrating agents are possible but concentrated sulfuric acid or heated aluminium oxide are most commonly used.

The reaction is elimination since a double bond is produced by the removal of OH from one carbon atom and H from an adjacent carbon atom.

▼ Study point

The dehydration of alcohols is basically the same reaction as the elimination of hydrogen halides from halogenoalkanes. Both reactions involve the formation of alkenes.

Classification of alcohols

Alcohols are **classified** as being primary, 1°, secondary, 2°, or tertiary, 3°, according to the bonding of the —OH in the molecule.

If the —OH is joined to a carbon that is itself joined to not more than one other carbon atom, the alcohol is primary.

If the —OH is joined to a carbon that is itself joined to two other carbon atoms, the alcohol is secondary.

If the —OH is joined to a carbon that is itself joined to three other carbon atoms, the alcohol is tertiary.

Examples of classification
(i) Methanol

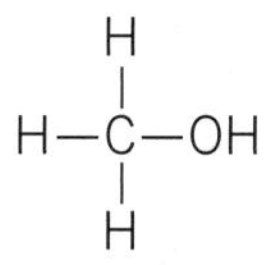

$$\begin{array}{c} H \\ | \\ H-C-OH \\ | \\ H \end{array}$$

The OH is joined to a carbon attached to no other carbons and so it is a primary alcohol.

(ii) Propan-1-ol

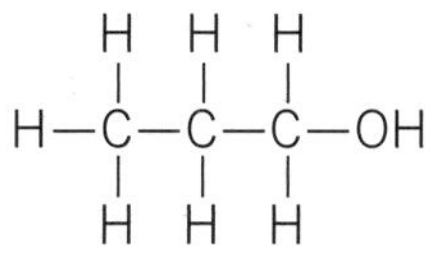

$$\begin{array}{ccc} H & H & H \\ | & | & | \\ H-C-C-C-OH \\ | & | & | \\ H & H & H \end{array}$$

The OH is joined to a carbon attached to one other carbon and so it is a primary alcohol.

(iii) Propan-2-ol

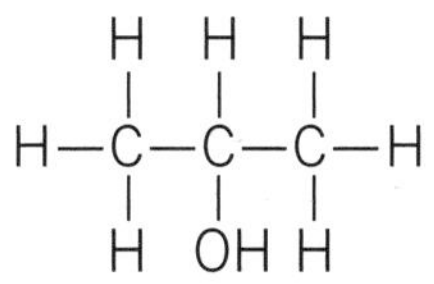

$$\begin{array}{ccc} H & H & H \\ | & | & | \\ H-C-C-C-H \\ | & | & | \\ H & OH & H \end{array}$$

The OH is joined to a carbon attached to two other carbons and so it is a secondary alcohol.

(iv) 2-methyl butan-2-ol

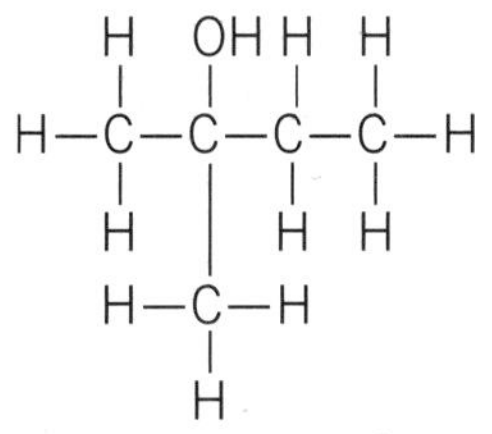

$$\begin{array}{cccc} H & OH & H & H \\ | & | & | & | \\ H-C-C-C-C-H \\ | & | & | & | \\ H & | & H & H \\ & H-C-H \\ & | \\ & H \end{array}$$

The OH is joined to a carbon attached to three other carbons and so it is a tertiary alcohol.

Key Term

Classification of alcohols classification of alcohols into primary, secondary or tertiary according to their structure.

Suggest the formulae of the two compounds that could be formed if ethane-1,2-diol, CH_2OHCH_2OH, was passed over heated aluminium oxide.

65

Knowledge check

Write the equation for the reaction that occurs when 2-methylpropan-1ol is passed over heated aluminium oxide.

66

Knowledge check

What happens when 2-methylbutan-2-ol is heated with concentrated sulfuric acid. Explain your answer.

67

Knowledge check

Classify the following as 1°, 2° or 3° alcohols.

(a) Aminomethanol, NH_2CH_2OH
(b) 2-methylbutan-2-ol, $CH_3C(CH_3)(OH)CH_2CH_3$
(c) Butan-2-ol $CH_3CH(OH)CH_2CH_3$
(d) Butan-1-ol, $CH_3CH_2CH_2CH_2OH$

Oxidation of alcohols

Acidified potassium dichromate(VI) can be used to oxidise many alcohols, Since dichromate(VI) will only behave satisfactorily as an oxidising agent in the presence of H^+ the oxidation is generally carried out by heating the alcohol with a mixture of aqueous potassium dichromate(VI) and sulfuric acid.

In equations showing oxidation of organic compounds the oxidising agent is usually shown as [O].

What happens under these conditions depends on whether the alcohol is primary, secondary or tertiary.

(i) Primary – using ethanol as an example

The reaction takes place in two stages.

Stage 1

$$H-\overset{\overset{\displaystyle H}{|}}{\underset{\underset{\displaystyle H}{|}}{C}}-\overset{\overset{\displaystyle H}{|}}{\underset{\underset{\displaystyle H}{|}}{C}}-O-H \quad + \quad [O] \quad \longrightarrow \quad H-\overset{\overset{\displaystyle H}{|}}{\underset{\underset{\displaystyle H}{|}}{C}}-C\underset{\diagdown H}{\overset{\diagup\!\!\diagup O}{}} \quad + \quad H_2O$$

Ethanol   Ethanal

Two hydrogen atoms are lost – one from the alcohol OH and one from the adjacent carbon. This creates a carbon to oxygen double bond.

The product, with the functional group

$$C\overset{\diagup\!\!\diagup O}{\underset{\diagdown H}{}}$$

is an aldehyde. In this case ethanal.

Stage 2

The aldehyde is oxidised further.

$$H-\overset{\overset{\displaystyle H}{|}}{\underset{\underset{\displaystyle H}{|}}{C}}-C\underset{\diagdown H}{\overset{\diagup\!\!\diagup O}{}} \quad + \quad [O] \quad \longrightarrow \quad H-\overset{\overset{\displaystyle H}{|}}{\underset{\underset{\displaystyle H}{|}}{C}}-C\underset{\diagdown OH}{\overset{\diagup\!\!\diagup O}{}}$$

Ethanal   Ethanoic acid

An oxygen is added to the aldehyde. The product, with the functional group

$$-C\underset{\diagdown O-H}{\overset{\diagup\!\!\diagup O}{}}$$

is a carboxylic acid. In this case ethanoic acid.

(ii) Secondary – using propan-2-ol as an example

The reaction has only one stage – this corresponds to stage 1 of the oxidation of primary alcohols.

$$H-\overset{\underset{|}{H}}{C}-\overset{\underset{|}{O}}{C}-\overset{\underset{|}{H}}{C}-H + [O] \longrightarrow H-\overset{\underset{|}{H}}{C}-\overset{\underset{\parallel}{O}}{C}-\overset{\underset{|}{H}}{C}-H + H_2O$$

Propan-2-ol

The product, with the functional group —C=O, is a ketone, in this case propanone.

(iii) Tertiary – using 2-methylpropan-2-ol as an example

$$CH_3-\overset{\underset{|}{\overset{|}{O}}{\underset{|}{H}}}{C}-CH_3$$
(with CH₃ below)

2-Methylpropan-2-ol

Since there is no hydrogen on the adjacent carbon that can be lost, no reaction takes place.

Summary of oxidation reactions
Primary alcohols ⟶ aldehydes ⟶ carboxylic acids

Seconday alcohols ⟶ ketones

Tertiary alcohols are not oxidised.

Use of acidified potassium dichromate(VI)
When it behaves as an oxidising agent acidified potassium dichromate(VI) changes colour from orange to green. This can be used as a test for primary and secondary alcohols since they will give a positive test result but tertiary alcohols will not.

Carboxylic acids

Carboxylic acids are an homologous series that contain the functional group $-C\overset{O}{\underset{O-H}{}}$.

This is often written as —COOH or —CO₂H. Many commonly used substances contain carboxylic acids and the acids they often have names that refer to these sources. The names used in scientific situations are, of course, based on the rules for naming organic compounds that are in Section 2.4. Some examples of commonly used and chemical names are given in the table.

Chemical name	Commonly used name	Formula	Present in	Reason for name
Methanoic acid	Formic acid	HCOOH	Venom from ant bites	Latin for ant is formica
Ethanoic acid	Acetic acid	CH₃COOH	Vinegar	
2-hydroxy-1,2,3-propane-tricarboxylic acid	Citric acid	HOOCC(OH)(COOH)COOH	Lemons and other fruits	Fruits are citrus
2-hydroxy-butane-1,4-dicarboxylic acid	Malic acid	HOOCCH₂CH(OH)COOH	Many fruits	Malus is Latin name for apple tree

Study point
In the oxidation equation for stage 1 you must not forget to balance the equation by including the water produced.

68
Knowledge check
What apparatus would you use if you wanted to prepare a carboxylic acid from an alcohol. Explain your answer.

YOU SHOULD KNOW ›››
››› the reactions of carboxylic acids

Key Term
Carboxylic acid An homologous series containing —COOH as the functional group.

Key Term

Weak acid An acid that is only partially ionised in aqueous solution.

Study point

Water is needed for the acid to produce H^+ ions but it is rarely included in the equilibrium equation.

Stretch & Challenge

To explain why H^+ ions are released you need to look at the structure.

Electron movement towards π bond

The C═O double bond is a π bond, similar to the C═C in alkenes. The lone pairs on the O are then also delocalised into this π bond. This pulls electron away from the O—H bond and weakens it. This makes it easier to break and release H^+ ions.

Study point

The salt produced is ionic. It can be written as $RCOO^-Na^+$ or the charges on the ions can be omitted. It **must** not be shown with a covalent bond to the Na.

69 Knowledge check

Write the equation, including state symbols, for the reaction between:

(a) Solid calcium carbonate and ethanoic acid.

(b) Aqueous potassium hydrogencarbonate and methanoic acid.

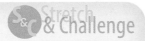

Structure of carboxylic acids

Using R as an alkyl group the structure is:

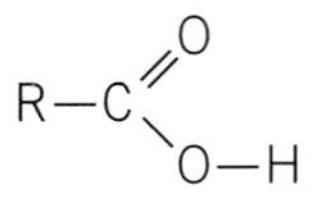

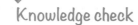

Carboxylic acids are acidic because they release H^+ ions when added to water. The pH of these solutions shows that they are **weak acids** so that the ionisation needed to produce H^+ is an equilibrium.

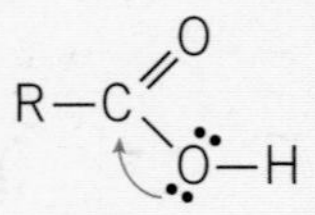

$$RCOOH \rightleftharpoons RCOO^- + H^+$$

Reactions of carboxylic acids

1 As an acid

Carboxylic acids react with bases, carbonates and hydrogencarbonates in a similar way to the reactions of inorganic, strong acids.

(i) Bases

Both soluble bases, i.e. alkalis and solid bases, such as metal oxides, neutralise carboxylic acids.

General equation Acid + Base ⟶ Salt + Water

Examples

The alkali sodium hydroxide

$$\underset{\text{ethanoic acid}}{CH_3COOH(aq)} + NaOH(aq) \longrightarrow \underset{\text{sodium ethanoate}}{CH_3COONa(aq)} + H_2O(l)$$

Since all reagents and products are colourless and soluble in water no **visible** change occurs.

The solid base copper(II) oxide.

$$\underset{\text{methanoic acid}}{2HCOOH(aq)} + CuO(s) \longrightarrow \underset{\text{copper(II) methanoate}}{(HCOO)_2Cu(aq)} + H_2O(l)$$

In this reaction, when warmed, the black solid, copper(II) oxide dissolves to form a blue solution of the salt.

(ii) Carbonates and hydrogencarbonates

Both carbonates and hydrogencarbonates react in a similar way.

General equation Acid + carbonate ⟶ salt + carbon dioxide + water

Examples

$$ZnCO_3(s) + 2C_2H_5COOH(aq) \longrightarrow (C_2H_5COO)_2Zn(aq) + CO_2(g) + H_2O(l)$$

$$NaHCO_3(aq) + CH_3COOH \longrightarrow CH_3COONa(aq) + CO_2(g) + H_2O(l)$$

Whether the carbonate or hydrogencarbonate is added to the aqueous acid as a solid or in aqueous solution carbon dioxide gas is produced. This means that an effervescence is seen and the gas can be shown to be carbon dioxide by testing it with lime water.

2 Esterification

Carboxylic acids react with alcohols:

General equation   Carboxylic acid + alcohol $\rightleftharpoons$ ester + water

This reaction is catalysed by the presence of concentrated sulfuric acid.

Example

$$CH_3COOH + C_2H_5OH \rightleftharpoons CH_3COOC_2H_5 + H_2O$$

ethanoic acid   ethanol   ethyl ethanoate

An easy way to obtain the correct formula for the ester is to draw the OHs of the acid and the alcohol next to each other. Then draw a box round the water to be removed and join the other parts of the molecules.

Example

The equation for the reaction between methanoic acid and propan-2-ol:

70

Knowledge check

(a) Write the equation that occurs when 2-chloro 3-methylbutanoic acid, $CH_3CH(CH_3)CHClCOOH$, reacts with butan-2-ol.

(b) Concentrated sulfuric acid is added to the reaction mixture. What is the purpose of the sufuric acid?

Esterification is therefore carried out by heating an alcohol with a carboxylic acid and concentrated sulfuric acid.

This reaction can be used as a test for an alcohol or a carboxylic acid since the esters produced have characteristic sweet, fruity odours.

If a pure sample of the ester is needed the carboxylic acid, the alcohol and concentrated sulfuric acid are heated together in a flask. This produces an equilibrium mixture of ester with some carboxylic acid and alcohol. In order to have a good yield of ester the flask is attached to allow the removal of the ester as soon as it is formed, i.e. to distil off the ester.

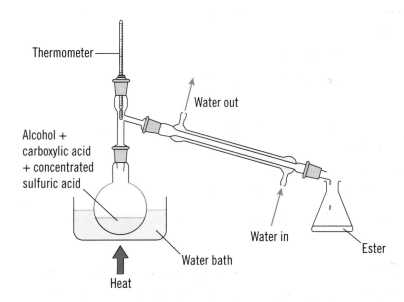

71

Knowledge check

Why can the ester be distilled off, leaving the other organic substances in the reaction flask?

Le Chatelier's principle then says that the removal of the ester pulls the equilibrium to the right and more ester is made.

Unit 2

2.8
Instrumental analysis

Traditionally chemists working in laboratories investigated samples of unknown substances using chemical tests. Nowadays, however, instrumental techniques are much more commonly used. These methods require very small samples to test and so are less damaging or invasive. They are also usually quick to carry out and they give very accurate data.

Different instrumental techniques give different information about a compound so that often more than one type of spectrum is used to determine the structure of an unknown.

Topic contents

You should be able to demonstrate and apply your knowledge and understanding of:

- The use of mass spectra in identification of chemical structure.
- The use of IR spectra in identification of chemical structure.
- The use of ^{13}C and low resolution ^{1}NMR spectra in identification of chemical structure.

Traditionally volumetric or gravimetric techniques were used to analyse the nature and quantity of an unknown substance. In volumetric analysis, titrations are used to measure volumes of solutions whilst in gravimetric analysis, masses of solids are found. Nowadays spectroscopic techniques are more often the preferred method.

In this unit some of the available techniques are considered.

Mass spectrometry

The main features of a mass spectrometer are in the diagram:

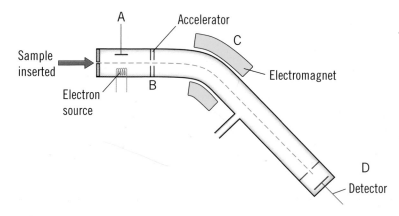

The sample is inserted and, at A in the mass spectrometer, high-energy electrons knock an electron off the molecules. This makes positive ions that are accelerated by B. The positive ions are then deflected by an electromagnet at C and are detected at D. The dotted line shows the path of a molecule through the spectrometer.

Although it is not necessary for you to be able to draw this diagram, understanding how a mass spectrum is produced is helpful.

When an electron is knocked off the molecule of an organic compound a positive ion is produced. This is the **molecular ion** and is often shown as M^+. Bombardment with the high energy electrons means that some molecules are split into smaller parts – **fragments.** The molecular ion and positively charged fragments are accelerated by the negatively charged plates and then deflected by the electromagnet. Those that pass through the slit at the end of the spectrometer are detected.

The amount of deflection for each positively charged species depends on the mass – the heavier the ion the less it is deflected. If the strength of the electromagnetic field is altered, species with different masses will pass through the slit and be detected.

This means that a mass spectrum can be produced.

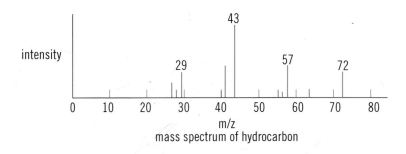

mass spectrum of hydrocarbon

YOU SHOULD KNOW ›››

››› the use of mass spectra in identification of chemical structure

››› the use of IR spectra in identification of chemical structure

››› the use of ^{13}C and low resolution ^1H spectra in identification of chemical structure

Key Terms

Molecular ion the positive ion formed in a mass spectrometer from the whole molecule.

Fragmentation splitting of molecules in a mass spectrometer into smaller parts.

! Extra Help

Imagine rolling a table tennis ball and a ten-pin bowling ball across a table. If someone blows from the side, the lighter table tennis ball will be deflected off its course much more than the heavier ten-pin bowling ball. This is the same as the lighter particles in a mass spectrometer being deflected more by the electromagnetic field.

The x-axis shows mass/charge (m/z) but you can assume that the charge on all the ions is 1 so that the x-axis shows the mass of the particles present. The y-axis shows the intensity of each peak, i.e. it is a measure of the abundance of each positive ion. Although this can be helpful in more advanced studies investigating the structure of compounds, it is not generally used.

In the mass spectrum the peak with the largest m/z is the molecule that has only lost an electron, i.e. the molecular ion, M^+. As the electron has a negligible mass the mass of this ion is the same as the mass of the molecule. This then gives the M_r.

The mass spectrum above is that of a hydrocarbon. The largest m/z is at 72 and therefore the M_r is 72. This value suggests that five carbon atoms are present (mass $5 \times 12 = 60$) and so 12 left, i.e. compound, i.e. C_5H_{12}. This means that it an isomer of pentane.

The fragments are then used to give information about which isomer it actually is. $29 = C_2H_5^+$,

$43 = CH_3CH_2CH_2^+$, $57 = CH_3CH_2CH_2CH_2^+$. This suggests that the compound is pentane $CH_3CH_2CH_2CH_2CH_3$.

In the mass spectrum of compounds containing chlorine or bromine more than one M_r peak will be produced. These, and other fragment peaks, correspond to the presence of more than one isotope for each halogen, i.e. ^{35}Cl and ^{37}Cl and ^{79}Br and ^{81}Br.

An example is a simplified version of the mass spectrum of 2-chloropropane.

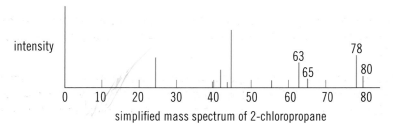

simplified mass spectrum of 2-chloropropane

In the spectrum the M^+ peak at 78 is due to $C_3H_7{}^{35}Cl$ and that at 80 due to $C_3H_7{}^{37}Cl$. The peaks at 63 and 65 are due to the loss of 15, i.e. CH_3^+ from the molecule. The other fragments are caused by rearrangements and are difficult to interpret from the structure of 2-chloropropane.

Infrared spectroscopy

Radiation in the IR part of the electromagnetic spectrum is absorbed to cause increased vibrations and bending in organic molecules. In an IR spectrometer a range of IR radiation of different energies is passed through the sample and the spectrum produced shows the energies that were absorbed. Since these energies are **characteristic** of the bonds present the **absorptions** can be used to identify which bonds are present. This means that the functional group can be identified.

In any question you will be given the **wavenumbers** that are needed for the spectrum given. Some examples are given in the table.

Bond	Wavenumber/cm^{-1}
C — C	650 to 800
C — O	1000 to 1300
C = O	1650 to 1750
C — H	2800 to 3100
O — H	2500 to 3550

The IR spectrum for ethanol is below.

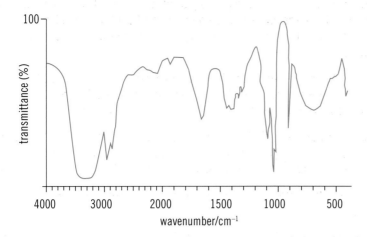

wavenumber/cm⁻¹ is labelled on the x-axis with values 4000, 3000, 2000, 1500, 1000, 500. The y-axis is labelled transmittance (%) with 100 at top.

Since peaks can be caused by many kinds of increased energy bending and stretching in the molecules, not all the peaks can be attributed to the presence of particular functional groups. This means that in interpreting an IR spectrum you should consider the information you have and look for absorptions corresponding to the functional groups/bonds present. The presence (or absence) of a peak at about $1700\,\text{cm}^{-1}$ is particularly useful since it is always sharp and clear. If seen it shows the presence of $C{=}O$.

The spectrum of ethanol is consistent with the structure of ethanol since it shows an absorption at approximately $1050\,\text{cm}^{-1}$ from $C{-}O$ and one at approximately $3000\,\text{cm}^{-1}$ from $O{-}H$.

It would be difficult to positively identify ethanol without some information from another source. The peak assigned to $O{-}H$ can cause confusion since all organic molecules contain $C{-}H$ bonds and the absorption due to this is in a very similar range of wavenumbers.

The IR spectra of complex molecules contain many peaks and it would be difficult to interpret these using individual wavenumber values. Data bases exist that have the IR spectra for a huge range of molecules and the **whole spectrum** of an unknown compound can be compared with these.

Nuclear magnetic resonance spectroscopy

Nuclear magnetic resonance (NMR) spectroscopy also corresponds to energy being absorbed to cause a change within molecules. In this case it is to reverse the spin of the **nucleus** of the atom within a **magnetic** field but you are not required to know anything about what actually happens when the energy is absorbed. The absorption of energy causes **resonance** and so it is called nuclear magnetic resonance. The energy absorbed is shown by the **chemical shift**, δ.

Different atoms in a molecule are joined to different atoms/groups and are thus said to be in different **environments.** The environment affects the energy needed to be absorbed to produce the change in the nucleus. This means that the absorption will appear at a different place on the spectrum, i.e. have different chemical shifts.

This technique is the basis of MRI (magnetic resonance imaging) scans carried out as a diagnostic tool in medicine. Perhaps the slight change in name is to avoid the inclusion of the term 'nuclear'. This might seem too close to being involved in a nuclear reactor for many patients!

▼ **Study point**

Questions will involve interpreting NMR spectra and will not require any knowledge of how they are produced.

Atomic nuclei spin and some nuclei have different energies, according to their direction of spin, when the molecule is placed in a magnetic field. This means that energy can be absorbed to promote the nucleus from the lower to the higher energy level. The change from lower to higher energy level is called resonance. To have these energy differences the atom must have an uneven number of nucleons so that commonly used forms of NMR spectroscopy are based on ^{13}C and ^{1}H.

▼ **Study point**

The height of each peak does **not** give any information that is useful in identifying the compound present.

^{13}C Spectroscopy

Naturally occurring carbon contains a very small percentage of the isotope ^{13}C. The presence of atoms of this isotope means that organic compounds will absorb energy and ^{13}C NMR spectra can be produced.

This spectrum will have chemical shifts, δ, on the x-axis and absorption on the y-axis.

An example of a ^{13}C spectrum is that for propan-1-ol.

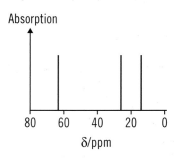

Data on δ values are needed to interpret the spectrum. This will be provided in any question. Some examples are given in the table.

Type of carbon	Chemical shift, δ/ppm
C—C	5 to 55
C—O	50 to 70
C—Cl	30 to 70
C=C	115 to 145
C=O	190 to 220

The spectrum gives two types of information:

- The **number** of different carbon environments – from the number of peaks.
- The **types** of carbon environment – from the chemical shifts.

From the spectrum there are three peaks and therefore C atoms in three environments.

Looking at the chemical shifts to find the environments:

- peak at δ = 64 ppm due to **C**—O
- peak at δ = 15 ppm due to **C**—C
- peak at δ = 27 due to **C**—C.

These can be seen in the structure of propan-1-ol.

$$H-\underset{H}{\overset{H}{\underset{|}{\overset{|}{C}}}}_1-\underset{H}{\overset{H}{\underset{|}{\overset{|}{C}}}}_2-\underset{H}{\overset{H}{\underset{|}{\overset{|}{C}}}}_3-O-H$$

It is not possible to tell, using the information available, which of the two C—C peaks is due to which atom involved in the C to C bond.

76

Knowledge check

Predict the number of peaks and the chemical shifts for each peak for:

(a) Ethanol, C_2H_5OH

(b) Butanone, $CH_3COCH_2CH_3$.

^{1}H spectroscopy

This can be called proton spectroscopy. It is similar to ^{13}C spectroscopy in that the spectra give information about:

- The **number** of different proton environments – from the number of peaks.
- The **types** of proton environments – from the chemical shifts.

However, a ^{1}H spectrum also gives the ratio of the **numbers of protons** in each environment. This ratio is shown by the relative area/ height of the peaks but is usually quoted in questions.

Typical values for δ are given in the table.

Type of proton	Chemical shift, δ/ppm
R—CH$_2$	0.7 to 1.6
R—OH	1.0 to 5.5
R—CH$_2$—R	1.2 to 1.4
R$_3$—CH	1.6 to 2.0
—C=OCH—	1.9 to 2.9
—O—CH$_3$	
—O—CH$_2$R	3.3 to 4.3
—O—CHR$_2$	
–CHO	9.1 to 10.1

Study point

The values for δ on the spectrum can differ considerably so do not be too concerned if your suggested structure would suggest δ values somewhat removed from those that the data sheet gives.

An example of a ^{1}H NMR spectrum is below.

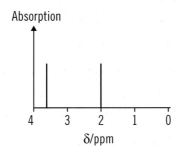

This spectrum is for a compound with molecular formula of $C_3H_6O_2$. The ratio of the area/ height of the peaks is 1:1. To identify the compound:

- From the molecular formula there are six protons.
- There a two peaks and therefore two proton environments.
- The ratio of the area/height of peaks is 1:1 so there must be three protons in each environment.
- Using the chemical shifts peak at δ = 3.6 matches O—C**H** and must be —C**H**$_3$
  peak at δ = 2.1 matches O=C—C**H** and must be O=C—C**H**$_3$.

Compound is therefore CH_3COOCH_3.

Knowledge check

77

For each compound below predict the number of peaks, the ratio of the peak areas and the chemical shifts in a proton NMR spectrum.

(a) $CH_3C=OCH_3$

(b) CH_3CH_2CHO

(c) $(CH_3)_2CHCH_3$

Exam practice questions

2.1

1 (a) For the energy cycle

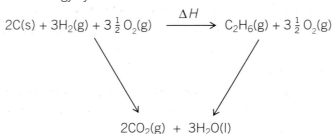

$$2C(s) + 3H_2(g) + 3\tfrac{1}{2}O_2(g) \xrightarrow{\Delta H} C_2H_6(g) + 3\tfrac{1}{2}O_2(g)$$

$$2CO_2(g) + 3H_2O(l)$$

Use the values in the table below to calculate the enthalpy change of reaction, ΔH.   [2]

Substance	Enthalpy change of combustion, $\Delta_c H^{\ominus}$ /kJ mol^{-1}
Carbon	−394
Hydrogen	−286
Ethane	−1560

(b) State what is meant by the term *standard molar enthalpy change of formation*.   [2]

(c) (i) Write an equation to represent the standard molar enthalpy change of formation, $\Delta_f H^{\ominus}$, of $H_2O(g)$ [1]

(ii) The standard molar enthalpy change of formation, $\Delta_f H^{\ominus}$, of $H_2O(g)$ is −242 kJ mol^{-1}. Using this value and the average bond enthalpies given in the table below, calculate the average bond enthalpy of the O—H bond in H_2O.   [2]

Bond	Average bond enthalpy /kJ mol^{-1}
H—H	436
O=O	496

2 Ethanol is an important industrial chemical and can be made by the direct hydration of ethene using a phosphoric acid catalyst.

$$CH_2{=}CH_2(g) + H_2O(g) \rightleftharpoons CH_3CH_2OH(g) \qquad \Delta H = -46\,kJ\,mol^{-1}$$

(a) Using the standard enthalpy change for the reaction above and the standard enthalpy change of formation, $\Delta_f H^{\ominus}$, given in the table below, calculate the standard enthalpy change of formation of gaseous ethanol.   [3]

Compound	$\Delta_f H^{\ominus}$ /kJ mol^{-1}
$CH_2{=}CH_2(g)$	52.3
$H_2O(g)$	−242

(b) Another way of calculating enthalpy change of reactions is by using average bond enthalpies.

Use the values in the table below to calculate the enthalpy change for the direct hydration of ethene.   [3]

Bond	Average bond enthalpy / kJ mol^{-1}
C—C	348
C=C	612
C—H	412
C—O	360
O—H	463

(c) (i) Give a reason why the calculated value in (b) is different to the actual value, −46 kJ mol^{-1}. [1]

(ii) Explain whether your answer to part (i) supports the use of average bond enthalpies to calculate the energy change for a reaction. [1]

(d) In some countries ethanol is replacing petrol (octane) as a car fuel.

(i) When ethanol, C_2H_5OH, is burnt in air, the only products are carbon dioxide and water.

Balance the equation for this reaction. [1]

$C_2H_5OH + O_2 \longrightarrow CO_2 + H_2O$

(ii) Use the standard enthalpy change of formation values given in the table below to calculate the standard enthalpy change, $\Delta_cH^\ominus$, for the combustion of ethanol. [2]

Compound	$\Delta_fH^\ominus$ / kJ mol^{-1}
$C_2H_5OH(l)$	−278
$CO_2(g)$	−394
$H_2O(l)$	−286
$O_2(g)$	0

(iii) The standard enthalpy change of combustion for octane $\Delta_cH^\ominus(C_8H_{18})$ is −5512 kJ mol^{-1}.

Using this value and your answer to (d)(ii) show that octane gives more energy per gram of fuel burned. [2]

(iv) Suggest a reason why ethanol is being used rather than petrol. [1]

3 Lisa was asked to measure the molar enthalpy change for the reaction between magnesium and copper(II) sulfate solution.

$Mg(s) + CuSO_4(aq) \longrightarrow MgSO_4(aq) + Cu(s)$

She was told to use the following method:

- Weigh out about 0.90 g of powdered magnesium.

- Accurately measure 50.0 cm^3 of copper(II) sulfate solution of concentration 0.500 mol dm^{-3} into a polystyrene cup (placed in another polystyrene cup to provide insulation).

- Place a 0.2 °C graduated thermometer in the solution and measure its temperature every half minute, stirring the solution before reading the temperature.

- At the third minute add 0.90 g of powdered magnesium, but do not record the temperature.

- Stir the mixture thoroughly, then record the temperature after three and a half minutes.

- Continue stirring and record the temperature at half-minute intervals for a further four minutes.

After carefully recording the temperature of the solution, she plotted a graph and found that the maximum temperature rise was 9.6 °C.

(a) Explain why the temperature of the copper(II) sulfate solution was measured for three minutes before adding the magnesium. [1]

(b) Calculate the heat given out during this experiment.

(Assume that the density of the solution is 1.00 g cm^{-3} and its specific heat capacity is 4.18 J K^{-1} g^{-1}) [2]

(c) Calculate the molar enthalpy change for the reaction between magnesium and copper(II) sulfate solution. [3]

(d) Name a piece of apparatus that Lisa could use to accurately measure 50.0 cm^3 of the solution. [1]

(e) State why she did not need to accurately weigh the powdered magnesium. [1]

(f) Explain why it is better to use powdered magnesium rather than a strip of magnesium ribbon. [2]

(g) The data book value for this molar enthalpy change is $-93.1\,kJ\,mol^{-1}$. Express the difference between Lisa's value and this value as a percentage of the data book value. [1]

(If you do not have an answer in (c) assume that the molar enthalpy change is $-65\,kJ\,mol^{-1}$, although this is not the correct answer.)

(h) State the **main** reason for Lisa's low value in this experiment and suggest one change that would improve her result. [2]

4 (a) In a neutralisation experiment, $50.0\,cm^3$ of aqueous sodium hydroxide solution of concentration $0.500\,mol\,dm^{-3}$ were added to $50.0\,cm^3$ of aqueous nitric acid solution of concentration $0.500\,mol\,dm^{-3}$ in a calorimeter. The initial temperatures of the solutions were $18.0\,°C$ and the temperature rose to $21.4\,°C$.

Assuming that the specific heat capacity of all the solutions is $4.18\,J\,g^{-1}\,K^{-1}$ calculate:

   (i) The number of moles of acid used. [1]

   (ii) The heat evolved in the experiment, in joules. [2]

   (iii) The enthalpy change of neutralisation for this reaction, in kilojoules per mole. [2]

(b) (i) State Hess's Law. [1]

   (ii) Explain why Hess's Law follows from the principle of conservation of energy. [1]

(c) Use the standard enthalpy change of formation values, $\Delta_f H^{\ominus}$, given in the table below to calculate the standard enthalpy change, $\Delta H^{\ominus}$, for the following reaction.

$$2N_2(g) + 2H_2O(g) + 5O_2(g) \longrightarrow 4HNO_3(l)$$ [2]

Species	$\Delta_f H^{\ominus}$ / kJ mol^{-1}
$N_2(g)$	0
$H_2O(g)$	−242
$O_2(g)$	0
$HNO_3(l)$	−176

5 (a) Ethene can be catalytically hydrogenated to ethane. The equation for the reaction is

$$C_2H_4(g) + H_2(g) \longrightarrow C_2H_6(g) \qquad \Delta H = -124\,kJ\,mol^{-1}$$

Calculate the average bond enthalpy, in $kJ\,mol^{-1}$, for the $C{=}C$ bond, by using the enthalpy change for the reaction and the average bond enthalpy values in the table below. [3]

Bond	Average bond enthalpy/kJ mol^{-1}
C—H	412
H—H	436
C—C	348

(b) Determine the value of ΔH, in $kJ\,mol^{-1}$, in the energy cycle below [1]

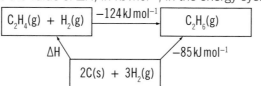

(c) Sulfur dioxide can be formed in the atmosphere, when hydrogen sulfide, emitted in gases from volcanoes, reacts with the air. The equation for the reaction is

$$2H_2S(g) + 3O_2(g) \longrightarrow 2H_2O(l) + 2SO_2(g)$$

(i) Use the standard enthalpy change of formation values given in the table to calculate the standard enthalpy change, $\Delta H^{\ominus}$, for the above reaction. [2]

Compound	$\Delta_f H^{\ominus}$ / kJ mol^{-1}
$H_2S(g)$	−20.2
$O_2(g)$	0
$H_2O(l)$	−286
$SO_2(g)$	−297

(ii) Explain why the standard enthalpy change of formation for $O_2(g)$ is zero. [1]

6 Iwan used the apparatus below to find the enthalpy change of combustion of nonane, C_9H_{20}.

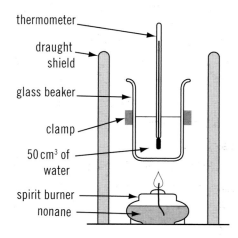

thermometer

draught shield

glass beaker

clamp

50 cm^3 of water

spirit burner

nonane

(a) Iwan measured the mass of the spirit burner at the start of the experiment and found that 0.20 g of nonane had been burned.

Calculate the number of moles of nonane present in 0.20 g. [2]

(b) The initial temperature of the water was 22.5 °C and the maximum temperature recorded during the experiment was 53.2 °C. Calculate the enthalpy change of combustion of nonane, in kJ mol^{-1}. Show your working. [3]

(The specific heat capacity, c, of water is 4.18 J g^{-1} K^{-1}).

(c) Give the main reason why the experimental value that Iwan obtained differs from the literature value. Suggest any improvements to the experiment that would give a more accurate value. [2]

2.2

1 Cobalt reacts with hydrochloric acid to give cobalt chloride and hydrogen.

$$Co(s) + 2HCl(aq) \longrightarrow CoCl_2(aq) + H_2(g)$$

    (a) Suggest a method for measuring the rate of this reaction. [1]

    (b) State what could be done to the cobalt to increase the rate of this reaction. [1]

2 **(a)** The diagram below shows the distribution of molecular energies for a sample of but-1-ene.

       On the diagram draw the distribution curve of molecular energies for the same sample of but-1-ene at a higher temperature. [1]

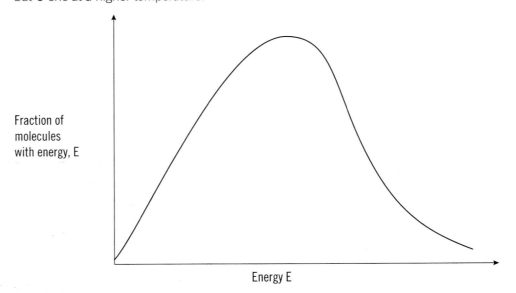

Fraction of molecules with energy, E

Energy E

    (b) But-1,3-diene dimerises slowly when it is heated to over 200°C as shown in the equation below.

$$2C_4H_6(g) \longrightarrow C_8H_{12}(g)$$

       The table below shows the initial concentration of but-1,3-diene and its concentration after 200 seconds.

Concentration of but-1,3-diene / mol dm^{-3}	Time / s
1.66×10^{-2}	0
1.60×10^{-2}	200

       (i) Use the values to calculate the initial rate of reaction in mol dm^{-3} s^{-1} [1]

       (ii) State how the value for the rate of reaction would change as the reaction proceeds at constant temperature. Give a reason for your answer. [2]

       (iii) The dimerisation of but-1,3-diene is an endothermic process.

          Sketch an energy profile for this reaction, clearly showing the activation energy, E_a. [2]

3 Ammonia is manufactured by the Haber process

$$N_2(g) + 3H_2(g) \rightleftharpoons 2NH_3(g) \qquad \Delta H = -92\,kJ\,mol^{-1}$$

State, giving reasons, what happens to the rate of the reaction as:

    (a) The temperature is increased. [2]

    (b) The total pressure is increased. [2]

4 (a) The decomposition of gaseous hydrogen iodide, HI, is represented by the equation below,

$$2HI(g) \rightleftharpoons H_2(g) + I_2(g) \qquad \Delta H = -53\,kJ\,mol^{-1}$$

Sketch and label on the axes below:

(i) the energy profile for the above decomposition. [1]

(ii) the energy profile for the same reaction if it were catalysed. [1]

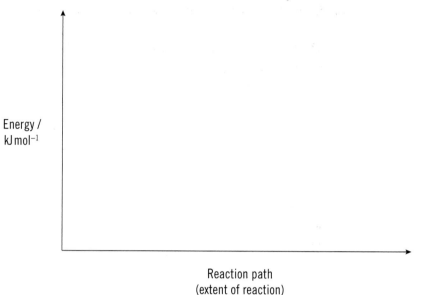

Energy /
kJ mol^{-1}

Reaction path
(extent of reaction)

(b) State how the collision theory explains the effect of changes in temperature on the rate of a reaction. [3]

5 (a) The rate of the reaction between dolomite, a mineral containing magnesium carbonate and calcium carbonate, and hydrochloric acid increases by a large amount if the temperature is increased.

Complete the following energy distribution curve diagram by drawing two lines that show the distribution of energies at two different temperatures.

Label the line at lower temperature T_1 and the line at higher temperature T_2. Use the diagram to help you explain why the rate increases as the temperature increases. [3]

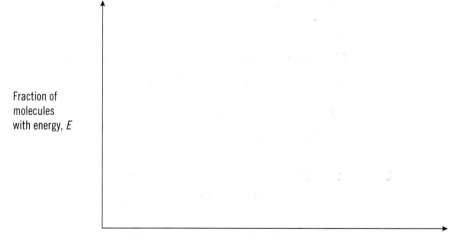

Fraction of
molecules
with energy, E

Energy E

(b) One method for following the rate of reaction is by measuring the volume of carbon dioxide evolved at suitable time intervals. Briefly outline a different method of following the rate of the reaction between dolomite and hydrochloric acid. [2]

6 Many catalysts are very expensive but their use does allow the chemical industry to operate more profitably. Explain why the use of catalysts provides economic and environmental benefits. [3]

7 Eurig is asked to measure the rate of reaction of calcium carbonate with dilute hydrochloric acid. He is given 1.50g of the carbonate and 10.0cm³ of acid of concentration 2.00 mol dm⁻³.

$$CaCO_3(s) + 2HCl(aq) \longrightarrow CaCl_2(aq) + CO_2(g) + H_2O(l)$$

(a) Give an observation that Eurig makes during this reaction. [1]

(b) Name a piece of apparatus that he could use to collect and measure the volume of carbon dioxide produced. [1]

(c) Suggest a method, other than measuring the amount of carbon dioxide produced at set time intervals, that Eurig could have used to follow the rate of this reaction. [1]

(d) Eurig repeats the experiment starting with a greater mass of calcium carbonate. He follows the rate of the reaction for 3 minutes.

He takes a number of measurements which include 150cm³ of carbon dioxide at 1 minute and 200cm³ at 2 minutes when the reaction finishes.

(i) Sketch a curve on a grid like that below to show these results. Label this graph **A**. [1]

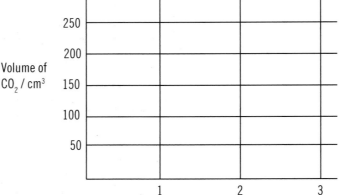

(ii) On the same grid sketch the graph that would be obtained if the experiment is repeated using hydrochloric acid of half the original concentration keeping all other factors the same. Label this graph **B**. [2]

(iii) Explain, using simple collision theory, why the rates of these two reactions are different. [2]

(e) With the aid of an energy distribution curve diagram, explain why raising the temperature by a small amount causes the rate of a chemical reaction to increase by a large amount. [3]

8 Carys carried out three experiments **A**, **B** and **C** to study the reaction between powdered magnesium and hydrochloric acid.

She used a gas syringe to measure the volume of hydrogen evolved at room temperature and pressure at set intervals. In each case, the amount of acid used was sufficient to react with all the magnesium.

$$Mg(s) + 2HCl(aq) \longrightarrow MgCl_2(aq) + H_2(g)$$

The details of each experiment are shown in Table 1 below.

Experiment	Mass of magnesium / g	Volume of HCl / cm³	Concentration of HCl / mol dm⁻³
A	0.061	40.0	0.50
B	0.101	40.0	1.00
C	0.101	20.0	2.00

▲ Table 1

The results obtained in experiment C are shown in Table 2 below.

Time / s	Volume of hydrogen / cm^3
0	0
20	50
40	75
60	88
80	92
100	100
120	100

▲ Table 2

(a) The results for experiments A and B have already been plotted on the grid below. On a copy of the same grid plot the results for experiment C and draw a line of best fit. [3]

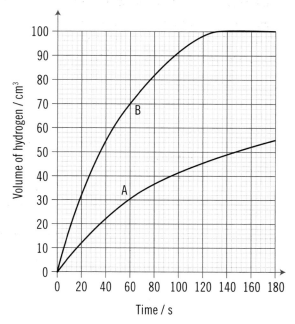

(b) (i) State in which experiment the reaction begins most rapidly and **use your graph** to explain your choice. [2]

(ii) By referring to Table 1 give an explanation of your answer in part (i). [1]

(c) State the volume of hydrogen evolved after 30 seconds in experiment **B**. [1]

(d) State one method of slowing down the reaction in experiment **C** and use collision theory to explain your choice. Assume that the quantities of magnesium and hydrochloric acid are the same as in Table 1. [3]

9 Measuring the rates of chemical reactions is very important in industrial processes, environmental studies and medical work.

(a) Name **three** factors that can affect the rate of a chemical reaction. [3]

(b) The following results were obtained in an experiment to find the rate of decomposition of hydrogen peroxide.

$$2H_2O_2 \longrightarrow 2H_2O + O_2$$

Time (s)	0	50	100	150	200	250	300
Volume O$_2$ (cm^3)	0	5.0	10.0	14.8	19.0	22.5	25.0

(i) Plot these results on graph paper and calculate the initial rate of reaction from your plot. **Show your working and state the units for the rate.** [5]

(ii) State how the rate of reaction changes over time and give a reason for any difference. [2]

(iii) Describe briefly how this experiment could be carried out. [2]

(c) Using collision theory for a reaction such as:

A(g) + B(g) $\longrightarrow$ C(g)

explain why the rate of reaction depends on both the pressure of the reactants and the temperature. [4]

10 The following results were obtained in an experiment to measure the rate of oxidation of iodide ions by hydrogen peroxide in acid solution as shown in the equation. The reaction was carried out at a temperature of 20°C.

$H_2O_2 + 2H^+ + 2I^- \longrightarrow I_2$(brown) + $2H_2O$

Time (s)	0	100	200	300	400	500
I_2 concentration (mol dm^{-3})	0	0.0115	0.0228	0.0347	0.0420	0.0509

(a) Plot these results on graph paper, labelling the axes and selecting a suitable scale. Draw the line of best fit. [3]

(b) Use the graph to calculate the initial rate of reaction and give the units. [2]

(c) Describe briefly the key features of the method that would have been used to obtain these results. [3]

(d) A similar experiment was carried out using hydrogen peroxide and iodide solutions of different concentrations. The initial rates calculated for each reaction are shown in the table.

Concentration of H_2O_2 (relative units)	Concentration of I$^-$ (relative units)	Initial rate (relative units)
0.60	0.050	4.1×10^{-4}
1.2	0.050	7.9×10^{-4}
1.2	0.10	1.6×10^{-3}

Analyse the data and state the relationship between the concentration of hydrogen peroxide and iodide ions and the initial rate of reaction. [2]

2.3

1 (a) Calculate the energies produced **per gram** when methane and butane are combusted.
ΔH(combn)methane = -890 kJ mol^{-1} ΔH(combn)butane = -2811 kJ^{-1} [2]

(b) Calculate the mass of carbon dioxide formed when **one gram** of methane and of butane are combusted. [2]

(c) From (a) and (b) state which would be the more environmentally friendly fuel for use in power stations, giving a reason. [2]

2 Motor vehicle environmental effects are judged on the grams of carbon dioxide emitted per kilometre of travel.

The combustion equations for bioethanol and ethane and the energy liberated are:

$C_2H_5OH + 3O_2 = 2CO_2 + 3H_2O$ $\qquad \Delta H = -1371$ kJ mol^{-1}

$C_2H_6 + \frac{7}{2}O_2 = 2CO_2 + 3H_2O$ $\qquad \Delta H = -1560$ kJ mol^{-1}

Bearing in mind that the car is driven by the energy liberated, state which fuel would be the more friendly per mol on the basis of these results, giving your reason.

There is, however, another factor to be considered. Discuss this and state whether or not your initial conclusion should be altered. [3]

3 Because of the link to global warming, much effort is being devoted to investigating how emissions of carbon dioxide, CO_2, into the atmosphere by power stations burning fossil fuels can be reduced or eliminated.

(a) One area of investigation is the removal of CO_2 by sodium carbonate. Thee possible reactions are:

$Na_2CO_3(s) + CO_2(g) + H_2O(l) = 2NaHCO_3(s)$ **1**

$3Na_2CO_3(s) + CO_2(g) + 5H_2O(l) = 2Na_2CO_3.NaHCO_3.2H_2O(s)$ **2**

$5Na_2CO_3(s) + 3CO_2(g) + 3H_2O(l) = 2Na_2CO_33NaHCO(s)$ **3**

(i) Giving a reason, determine from the equations which of the three reactions uses sodium carbonate, $Na_2CO_3(s)$, most effectively to absorb $CO_2(g)$. [2]

(ii) State Le Chatelier's Principle. [1]

(iii) Giving your reasons, use Le Chatelier's Principle to determine whether $CO_2(g)$ removal will be more efficient at high gas pressure or low gas pressure. [2]

(b) Another area of investigation is the use of a new type of plastic membrane, structured by means of nanotechnology, to catch carbon dioxide gas whilst allowing other waste gases to pass freely through.

If $1000\,dm^3$ of waste gas at $25°C$ yielded $275\,g$ of carbon dioxide, separated by a plastic membrane, calculate:

(i) The number of moles of carbon dioxide in the $275\,g$ separated by the membrane. [1]

(ii) The volume of carbon dioxide separated at $25°C$. [1]

[One mole of gas has a volume of $24.0\,dm^3$ at $25°C$ and $1\,atm$ pressure.]

(iii) The percentage by volume of carbon dioxide in the waste gas. [1]

(c) Carbon dioxide, CO_2 is an acid gas.

(i) Define the term acid. [1]

(ii) By considering its interaction with water, explain how carbon dioxide can behave as an acid. [1]

(iii) Though the pH of pure water is 7, explain why naturally occurring water in contact with air has a pH of less than 7. [1]

4 (a) The vast majority of motor vehicles worldwide are powered by petrol or diesel, which come from crude oil. Give **two** reasons why we cannot rely indefinitely on oil as a source of transport fuel. [2]

(b) Many vehicle manufacturers around the world have made the development of alternative fuels a priority. One such fuel being studied is hydrogen.

Its main advantage is that the only waste product is water; however, hydrogen does not occur naturally on Earth. It can be produced by passing an electric current through water.

(i) A leading car manufacturer said, 'Cars powered by hydrogen will be pollution-free'. Give **two** reasons why this is not necessarily true. [2]

(ii) A spokesperson for a safety group said, 'Hydrogen can burn explosively. It must not be used in cars unless it is 100 % safe'.

Comment on this. [1]

(c) Describe how industry is adapting to the challenges of Green Chemistry. Your answer should include reference to:

1   The overall aims of Green Chemistry.

2   Materials used or produced.

3   Energy used. [5]

5 Hydrocarbons play an important role in our life today, both as fuels and as raw materials for the synthesis of a wide range of materials. Most hydrocarbons are isolated from crude oil; however, there is increasing interest in alternative methods of obtaining these molecules.

One route to the production of hydrocarbons is the Fischer–Tropsch process, which uses hydrogen and carbon monoxide as starting materials to produce a range of molecules. The equation below shows the production of pentane, C_5H_{12}, by this route.

$$11H_2(g) + 5CO(g) = C_5H_{12}(l) + 5H_2O(l) \qquad \Delta H = -1049 \, kJ \, mol^{-1}$$

The enthalpies of formation of some of these substances are given in the table below:

Substance	Standard enthalpy of formation, $\Delta_f H^\ominus$ / kJ mol^{-1}
Hydrogen, $H_2(g)$	0
Carbon monoxide, $CO(g)$	−111
Water, $H_2O(l)$	−286

(a) (i)   State the temperature and pressure used as standard conditions. Give units for each.   [2]

(ii)   State why the standard enthalpy of formation for hydrogen gas is 0 kJ mol^{-1}.   [1]

(iii)   Use the values given to calculate the standard enthalpy of formation for pentane   [2]

(b) One method of producing the hydrogen gas required for the Fischer–Tropsch process is to use the reversible reaction below.

$$CO(g) + H_2O(g) = CO_2(g) + H_2(g) \qquad \Delta H = -42 \, kJ \, mol^{-1}$$

(i)   State and explain the effect, if any, of increasing pressure on the yield of hydrogen gas produced at equilibrium.   [2]

(ii)   State and explain the effect, if any, of increasing temperature on the yield of hydrogen gas produced at equilibrium.   [2]

(iii)   This reaction uses a catalyst based on iron oxide. State the effect of using a catalyst on the position of equilibrium.   [1]

2.4

1 (a) State the meaning of the term *heterolytic fission*.   [1]

(b) Complete the equation below to show the products of the heterolytic fission of the C—Cl bond in 2-methyl-2-chloropropane.   [1]

$$H_3C-\underset{\underset{CH_3}{|}}{\overset{\overset{CH_3}{|}}{C}}-Cl \longrightarrow$$

2 The compound below has a cherry odour and is used in the manufacture of fragrance agents.

$$\underset{H}{\overset{CH_3CH_2}{\diagdown}}C=C\underset{H}{\overset{CH_2OH}{\diagup}}$$

(a) Name the functional groups present in this compound.   [2]

(b) State the molecular formula of the compound.   [1]

3 Compounds **A** and **B** are organic compounds of sulfur found naturally in some foods.

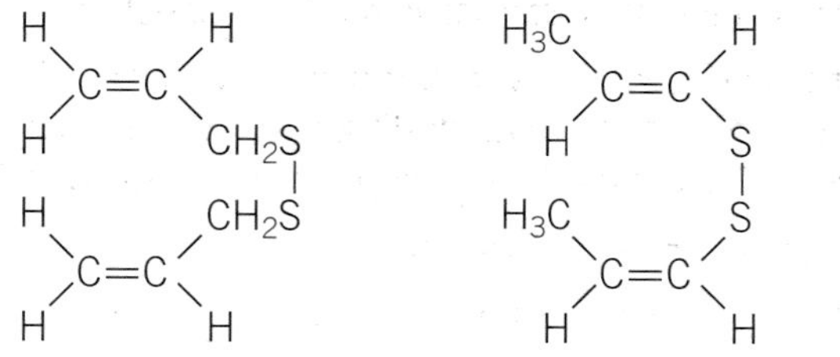

    Compond A   Compond B produced
    found in garlic   by cooking onions

(a) These two compounds are structural isomers. State what is meant by the term structural isomer. [1]

(b) Explain why only compound B can exist as *E-Z* isomers. [2]

(c) Compound A is sold by the chemical suppliers at £48.00 for 100g. The material sold is 73% pure but this is satisfactory for the purposes needed.

Calculate the cost of 1 mol of compound **A**, which has a molecular formula of $C_6H_{10}S_2$. [2]

4 A section of an addition polymer is shown below.

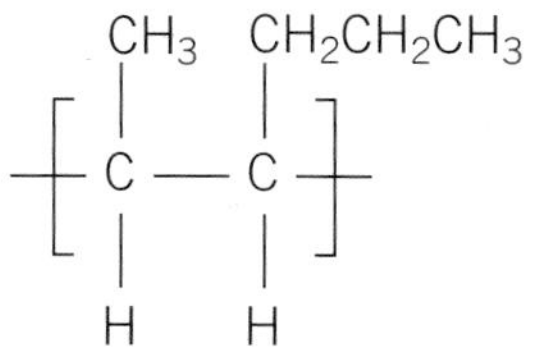

State the systematic name of the monomer that gives this polymer. [1]

2.5

1 (a) The decomposition of hydrogen peroxide, H_2O_2, may involve hydroxyl radicals.

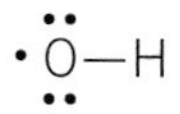

$$\cdot \overset{\cdot\cdot}{\underset{\cdot\cdot}{O}} - H$$

State why this is described as a radical. [1]

(b) Another reaction that produces radicals is the reaction of chlorine with propane.

    (i) Give the equation for the reaction of a propyl radical and chlorine gas. [1]

    (ii) Why is the reaction in (i) described as a propagation reaction? [1]

(c) Radicals are involved in the cracking of petroleum fractions at 600 °C.

One of the products obtained by cracking is an alkane of molar mass 100g. Deduce the molecular formula of this alkane. [1]

(d) Radicals are produced by the homolytic fission of a covalent bond. What is meant by homolytic bond fission? [1]

2 Petroleum is a mixture of saturated hydrocarbons, some of which are structural isomers of one another. These are separated into fractions by distillation. Some of these fractions are used to make important chemicals such as ethene while others are used to make fuels.

(a) Explain what is meant by a saturated hydrocarbon. [1]

(b) Propane and heptane are two of the hydrocarbons obtained from petroleum.

   (i) Write the equation for the complete combustion of propane. [1]

   (ii) 3-Ethylpentane is a structural isomer of heptane. Draw the skeletal formula of this isomer. [1]

(c) Describe the structure and bonding in an ethene molecule. [3]

(d) Ethene can be used as the starting material in the industrial preparation of ethanol but ethanol can also be made by the fermentation of glucose.

$$C_6H_{12}O_6 \longrightarrow 2C_2H_5OH + 2CO_2$$

Calculate the minimum mass of glucose required to give 230 g of ethanol. [3]

3 Crude oil is a complex mixture of hydrocarbons, with samples from different locations in the world having different compositions. The table blow gives the composition of crude oil from two locations.

Fraction	Percentage by mass	
	Brent crude	Gulf of Suez
petroleum gases	2.4	1.2
naphtha	19.1	13.6
kerosine	14.2	12.7
gas oil	20.9	18.7
residue	43.4	53.8

(a) The different fractions are separated by fractional distillation. Explain why the different fractions have different boiling temperatures. [2]

(b) The petroleum gases produced from crude oil can contain both propane and butane.

A barrel of Gulf of Suez crude oil has a mass of 145 kg. Assuming all the petroleum gas released from the oil is butane, calculate the volume that this gas would occupy at 1 atmosphere pressure. [3]

(c) Naphtha is used as the starting material for the production of other compounds. Under suitable conditions it can be cracked to produce ethene and alkanes. The ethene is then used to produce (poly)ethene. Discuss how poly(ethene) is produced from naphtha, by responding to the bullet points below.

- An explanation of which of the two types of crude oil given would be more useful for producing alkenes, such as ethene.

- An equation for the cracking of decane, giving ethene as one of the products.

- An explanation of what is meant by polymerisation.

- An equation for the polymerisation of ethene, clearly stating the type of polymerisation that is occurring.

- State a different polymer that is in common use, with the structure of the monomer used in its production [5]

2.6

1 (a) State the molecular formula of compound L that has the skeletal formula shown. [1]

(b) Compound L reacts with alcoholic sodium hydroxide solution to give hex-1,3-diene as one of the products.

State the type of reaction that has occurred [1]

2 One compound previously used in correction fluid was 1,1,1-trichloroethane, but this has been replaced by compounds such as methylcyclohexane, that has a much less adverse effect on the environment.

1,1,1–trichloroethane   Methylcyclohexane

(a) Explain, in terms of bond strengths, why 1,1,1-trichloroethane has an effect on the ozone layer but methylcyclohexane does not. [2]

(b) Hept-1-ene is an isomer of methylcyclohexane. Describe a chemical test that gives a positive result for hept-1-ene but not for methylcyclohexane. [2]

3 It is possible to test for the presence of halogen atoms in a halogenoalkane by hydrolysing the molecule and testing for the halide ion released, using silver nitrate solution.

The hydrolysis of three compounds was performed under identical conditions and the time taken for a precipitate to form was measured. The results were as follows:

Compound	Time for precipitate to form/ minutes
1-chloropropane	17
1-bromopropane	4
1-iodopropane	Less than 1

The carbon–halogen bond energies and the electronegativity differences for each bond are given below:

Bond	Average bond enthalpy/ kJ mol^{-1}	Electronegativity difference
C—Cl	338	0.61
C—Br	276	0.41
C—I	238	0.11

Use both tables to comment on the factors that affect the rate of reaction. [6]

2.7

1 **(a)** 1-Bromobutane is a liquid that is insoluble in water. It can be converted to butan-1-ol in a one-step reaction.

(i) Give the reagent(s) and conditions needed for this reaction. [2]

(ii) Explain why butan-1-ol is more soluble in water than 1-bromobutane. [3]

(b) Butan-1-ol can be converted into liquid butanoic acid in a one-step reaction.

(i) Give the reagent(s) and condition(s) required for this reaction. [2]

(ii) The reaction above frequently produces a mixture containing unreacted butan-1-ol and butanoic acid. State how these liquids could be separated. [1]

2 Alkenes can be precursors to many organic compounds as shown below.

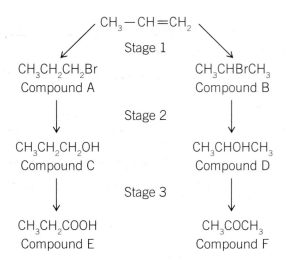

(a) (i) Draw the mechanism for the production of the major product in stage 1. In your answer you should classify the type of mechanism occurring. [5]

(ii) Explain why two products are formed in stage 1 and why one of these products is favoured. [2]

(b) In stage 3 a third product was also formed which is not shown above, This compound also has three carbon atoms and its NMR spectrum included a peak at chemical shift $\delta = 9.8$ ppm. Its IR spectrum included a peak at $1715\,cm^{-1}$.

Use the information above to deduce whether this compound is the result of the reaction of compound **C** or compound **D**. In your answer you should also include the type of reaction occurring in stage 3. [4]

(c) As part of a project, two students were asked to report on tests for functional groups present in the compound below:

$$H-\overset{\overset{\displaystyle H}{|}}{\underset{\underset{\displaystyle H}{|}}{C}}-\overset{\overset{\displaystyle H}{|}}{\underset{\underset{\displaystyle H}{|}}{C}}=\overset{\overset{\displaystyle H}{|}}{C}-\overset{\overset{\displaystyle H}{|}}{\underset{\underset{\displaystyle H}{|}}{C}}-Br$$

Nia reported that in testing for the C=C group:

- Aqueous bromine should be added to the compound and if the test is positive, the colour changes from purple to colourless.

- The name of the compound formed in this test is 2,3,4-tribromobutane.

(i) Correct the **two** mistakes in her report. [2]

(ii) David reported that in testing for bromine in the compound:

- Dilute hydrochloric acid should be added to the compound.

- Aqueous silver nitrate should then be added.

- You should see a cream precipitate.

State and explain the observations David would have made if he had carried out this test. [1]

Outline the correct method for carrying out a test for bromine in the compound. Include any reagents used in your answer. [2]

2.8

1 **(a)** Propene reacts with hydrogen bromide to give 2-bromopropane.

   (i) Draw the mechanism for this reaction. [3]

   (ii) Explain why the product of this reaction is mainly 2-bromopropane rather than 1-bromopropane. [2]

(b) Compound **C** is a compound of carbon, hydrogen and bromine only. Bromine has two isotopes, ^{79}Br and ^{81}Br, in equal abundance. Use all the information below to deduce the structure of compound **C**, giving your reasoning.

   • Compound **C** contains 29.8% carbon, 4.2% hydrogen and 66.0% bromine by mass.

   • The mass spectrum of compound **C** contains peaks at m/z of 15, 41 and a pair of peaks at 120 and 122.

   • The infrared spectrum of compound **C** has absorptions at $550\,cm^{-1}$, $1630\,cm^{-1}$ and $3030\,cm^{-1}$.

   • Compound **C** is a *Z*-isomer. [6]

2 Ethanoic acid, CH_3COOH, is an organic acid that gives vinegar its sour taste and pungent smell.

(a) The infrared spectrum of ethanoic acid is given. Label this spectrum to show the characteristic absorptions for three bonds present in ethanoic acid. [2]

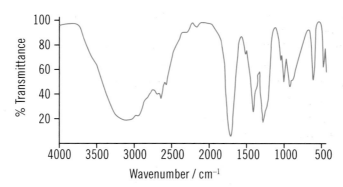

(b) The mass spectrum of ethanoic acid is shown below.

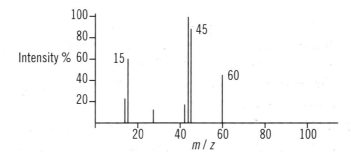

Explain how this spectrum shows that the formula of ethanoic acid is CH_3COOH. [2]

(c) Ethanoic acid can be formed from the oxidation of ethanol.

   (i) State a suitable oxidising agent and the conditions needed to carry out this oxidation. [1]

   (ii) What would be observed during this oxidation? [1]

(d) The boiling temperature of ethanol is 78 °C. Giving a reason in both cases, state how you would expect the boiling temperatures of the following compounds to differ from that of ethanol.

   (i) Propane

   (ii) Butan-1-ol [2]

Answers to knowledge check questions

Unit 1

1.1

1 **(a)** $P = 4, O = 10$

    (b) $Al = 2, O = 6, H = 6$

2 27

3 **(a)** Sodium sulfate

    (b) Calcium hydrogencarbonate

    (c) Copper(II) chloride

4 **(a)** Al_2O_3

    (b) K_2CO_3

    (c) $(NH_4)_2SO_4$

5 **(a)** −3

    (b) 0

    (c) +7

    (d) +6

6 **(a)** $2SO_2 + O_2 \longrightarrow 2SO_3$

    (b) $Fe_2O_3 + 3CO \longrightarrow 2Fe + 3CO_2$

7 $Ba^{2+}(aq) + SO_4^{2-}(aq) \longrightarrow BaSO_4(s)$

1.2

8 Cu-63: 29 p, 29 e, 34 n

    Cu-65: 29 p, 29 e, 36 n

9 **(a)** 53 p, 54 e

    (b) 12 p, 10 e

10 ^{234}Pa

11 24 days

12 **(a)** **(i)**

       (ii)

    (b) $1s^2 2s^2 2p^6 3s^2 3p^6 3d^{10} 4s^1$

13 9 (1 s, 3 p and 5 d)

14 The element belongs to group **2** in the periodic table because there is a **large jump** between the **second** and **third** ionisation energies.

15 **(a)** Line with 690 THz has the higher energy since $E \propto f$

    (b) Line with 460 THz has the higher wavelength since $f \propto 1/\lambda$

16 C

17 495 kJ mol^{-1}

1.3

18 248.3

19 1.008

20 The molecular ion, H_2^+, is not stable and splits to give a hydrogen atom and an H^+ ion.

21 **(a)** 23.0 g

    (b) 1533 g

22 **(a)** 2.12 g

    (b) 0.0136 mol

23 $FeCl_3$

24 0.816 dm^3

25 Pressure in Pa

    Volume in m^3

    Temperature in K

26 3.15 dm^3

27 338 cm^3

28 2.50×10^{-3} mol dm^{-3}

29 0.148 mol dm^{-3}

30 45.8%

31 **(a)** 4

    (b) 4

    (c) 1

    (d) 3

    (e) 5

1.4

32 **(i)** ionic, covalent, coordinate

    (ii) ionic, covalent, metallic, covalent, ionic.

33 MgO

34 A and D are intermolecular,

    B and C are intramolecular.

35 A 3, B 2, C 1

36 **(i)** tetrahedral, tetrahedral, linear, trigonal planar, trigonal bipyramid.

    (ii) 120°, 180°, 109°, 90°, 90/120° in list order.

1.6

37 **(i)** Decrease, increase

    (ii) Increases, decreases.

38 Electrons, electron. reducing, oxidised, reduced, in order.

39 (i) Fe 3, Cl –1, H 1, O –1, K 1, Cr 6, O –2, Br 0, H 1, C 4, O –2.

(ii) LHS Na 1, I –1, H 1, S 6, O –2,

RHS Na 1, S 6, O –2, I 0, H 1, S–2, H 1, O –2.

40 (b) and (c).

41 (a) **No reaction.**

(b) Bromine is liberated.

42 Copper(II) carbonate and hydrochloric acid

1.7

43 (a) No effect. There are 2 moles of gas on each side of the equation.

(b) Moves to the right. Moves in the endothermic direction. The reaction is endothermic from left to right.

44 (a) $K_c = \dfrac{[SO_3]^2}{[SO_2]^2\,[O_2]}\,dm^3\,mol^{-1}$

(b) $K_c = \dfrac{[PCl_3]\,[Cl_2]}{[PCl_5]}\,mol\,dm^{-3}$

45 50

46 Nitric acid fully dissociates in aqueous solution. The aqueous hydrogen ion concentration is equal in magnitude to the concentration of the acid.

47 A strong acid is one that fully dissociates in aqueous solution while a concentrated acid is one that consists of a large quantity of acid and a small quantity of water.

48 (a) 2

(b) $3.16 \times 10^{-3}\,mol\,dm^{-3}$

49 Potassium nitrate and water

50 It is not stable (in air) since it reacts with atmospheric carbon dioxide and it has a low molar mass.

51 (a) A measuring cylinder is not accurate enough so the exact amount of moles of base would not be known / the percentage error from the equipment would be too large

(b) To show the end point / when to stop adding acid

Unit 2

2.1

1 This is the enthalpy change when one mole of a substance is formed from its constituent elements in their standard states under standard conditions.

2 Oxygen gas is an element in its standard state.

3 $-235\,kJ\,mol^{-1}$

4 $-313\,kJ\,mol^{-1}$

5 $-484\,kJ\,mol^{-1}$

6 $83.6\,kJ$

7 $-54.3\,kJ\,mol^{-1}$

8 (a) Rate of reaction is quicker.

(b) Extrapolation gives the temperature that would have been reached if the reaction occurred instantly / extrapolation allows for heat loss during the experiment.

2.2

9 $0.004\,mol\,dm^{-3}\,s^{-1}$

10 There is an increase in concentration of acid therefore there are more molecules in a given volume, so there is an increase in the number of collisions per unit time. This means that there is a greater chance that the number of effective collisions will increase.

11 $-52\,kJ\,mol^{-1}$

12 Only the molecules with an energy equal to or greater than the activation energy are able to react. At the higher temperature, the mean kinetic energy of the acid molecules increases and many more molecules have sufficient energy to react.

13 A catalyst is a substance that increases the rate of a chemical reaction without being used up in the process. It increases the rate of reaction by providing an alternative route of lower activation energy.

14 Measure the volume of carbon dioxide produced (using a gas syringe) at constant time intervals / measure the change in mass at various times (using weighing scales).

15 Some of the hydrogen gas will escape before the bung is properly replaced.

16 (a) The total volume affects the concentrations of reactants and must remain the same to make a fair comparison between runs.

(b) Rates vary rapidly with changes in temperature.

2.3

17 Hydrogen

2.4

18 The longest carbon chain has 8 carbon atoms. Name based on 'oct'.

19 (a)

(b)

20 **(a)** 3-methyl pent-1-ene.

(b) 2-bromo, 2-chloropropan-1-ol.

21 **(a)** C_4H_7OCl

(b)

(c)

22 Alkenes have similar **chemical** properties but show a trend in **physical** properties because they all belong to the same **homologous series**.

23 $C_{100}H_{202}$.

24 Ratio number of moles $C:H:O = \dfrac{40.00}{12.0} : \dfrac{6.67}{1.01} : \dfrac{53.33}{16.0}$

$= 3.33 : 6.67 : 3.33$

Divide by smallest $= 1:2:1$

Empirical formula $= CH_2O$

Relative empirical mass $= 12 + 2 + 16 = 30$

$M_r = 55$ Molecular formula $= C_2H_4O_2$

25 0.660g of CO_2 contain $0.660 \times \dfrac{12}{44}$ g of carbon $= 0.18$g.

0.225g of H_2O contain $0.225 \times \dfrac{2}{18}$ g of hydrogen $= 0.025$g

Ratio $C:H = \dfrac{0.180}{12.0} : \dfrac{0.025}{1.01}$

$= 0.015 : 0.0248$

Divide by smaller $= 1 : 1.65$

$= 3:5$

Empirical formula $= C_3H_5$

Relative empirical mass $= 41$.

M_r given as approximately 80. Molecular formula $= C_6H_{10}$.

26

Pentane

2-Methylbutane

2,2-Dimethylpropane

27

Pent-1-ene   Pent-2-ene

28 A and B are isomers of each other.

29

30 Z isomer. Cl has a greater A_r than C.

31 **(a)**

(b) They are the same compound.

32

Intramolecular force

Intermolecular force

33 Any temperature in a range above 36°C would be acceptable. Suggest something between 50°C and 70°C.

2.5

34 Two of water, carbon dioxide and methane.

35 The presence of **nitric** acid and **sulfuric** acid in rain water **lowers** the pH.

36 $CaCO_3(s) + 2HNO_3(aq) \longrightarrow Ca(NO_3)_2(aq) + CO_2(g) + H_2O(l)$

37 The smoke is carbon – this forms from incomplete combustion of fuel.

38 $C_4H_{10} + 4\frac{1}{2}O_2 \longrightarrow 4CO + 5H_2O$

39 $C_8H_{18} + 12\frac{1}{2}O_2 \longrightarrow 8CO_2 + 9H_2O$

40 Alkanes are generally unreactive because **they are non polar and do not contain multiple bonds.** Radicals are very reactive because they **have an unpaired electron.**

41 $CH_4 + 4Cl_2 \longrightarrow CCl_4 + 4HCl$

42 Steamy fumes are seen because hydrogen chloride is formed.

43

44 Since they have to be manufactured from alkanes it is more cost effective to use the alkanes directly as fuels.

45 **1** Bromine water is brown and therefore easier to show loss of colour.

2 Bromine water is a liquid and easier to handle than the gas chlorine.

46 Nickel is cheaper. Although catalysts should not be changed when they are used, eventually they become 'poisoned' by impurities in the hydrocarbons and then are less efficient.

47 $(CH_3)_2C\!=\!CH_2 + HBr \longrightarrow (CH_3)_2CBrCH_3$.

48 Poly(1-bromo, 2-nitroethene).

49

50 CH_2.

51

2.6

52 2-methyl, 2-halogenobutane and sodium hydroxide.

53 Aqueous sodium hydroxide and heat.

54

55 Chlorine is smaller than bromine. The nucleus therefore exerts more attraction on electrons.

56 Chlorine is smaller than bromine. The chlorine atom can get closer to the carbon atom and therefore the attraction between the shared pair of electrons and the nuclei is greater.

57 Warm the suspected bromocompound with aqueous sodium hydroxide. Neutralise the excess sodium hydroxide with dilute nitric acid. Add aqueous silver nitrate. If the compound is a bromoalkane a cream precipitate that is soluble in concentrated ammonia is formed.

58 $Ag^+(aq) + I^-(aq) \longrightarrow AgI(s)$

59 **(a)** 2-methyl butan-1-ol.

(b) 2-methyl but-1-ene.

60 **(a)** 2-methyl butan-2-ol.

(b) 2-methyl but-1-ene and 2-methyl but-2-ene.

61 To form a radical a bond must be broken homolytically. The C—Cl bond is the weakest and therefore breaks most readily.

62 Fluorine is smaller than chlorine. The fluorine atom can get closer to the carbon atom and therefore the attraction between the shared pair of electrons and the nuclei is greater.

2.7

63 Ethanol could be removed by cooling the gaseous mixture. Ethanol and steam will liquefy. The liquid mixture can be separated by fractional distillation.

64 Fractional distillation is needed as the boiling temperatures of water and ethanol are quite close together.

65

66

'The H from either side of the C can be lost with the OH'.

67 **(a)** Primary.

(b) Tertiary.

(c) Secondary.

(d) Primary.

68 Apparatus suitable for reflux. You need to be able to heat the liquids for a time without losing them through evaporation.

69 **(a)** $CaCO_3(s) + 2CH_3COOH(aq)$
$\longrightarrow (CH_3COO)_2Ca(aq) + CO_2(g) + H_2O(l)$

(b) $KHCO_3(aq) + HCOOH(aq)$
$\longrightarrow HCOOK(aq) + CO_2(g) + H_2O(l)$

70 (a)

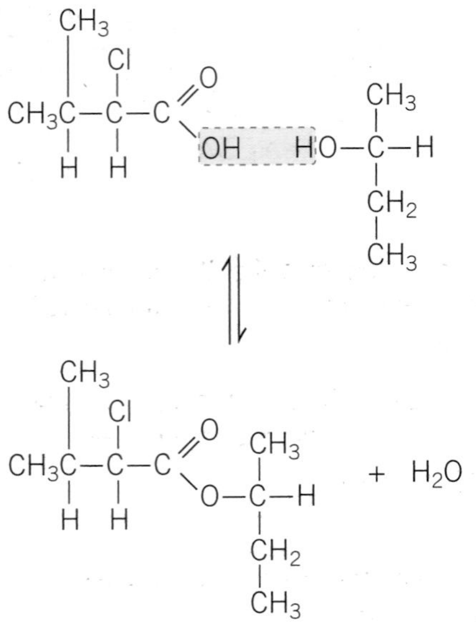

(b) Sulfuric acid acts as a catalyst.

71 The ester has the lowest boiling temperature. Alcohols, carboxylic acids and water all contain the —OH group and can hydrogen bond. This increases the boiling temperatures. Esters cannot hydrogen bond.

2.8

72 In a mass spectrometer **positive** ions are produced when an **electron** is knocked off. These ions are deflected by a **magnetic field** with the **heaviest** being deflected least. The M_r of the compound is shown, in the mass spectrum, by **the peak with the largest m/z.**

73 (a) 60. This is the molecular ion and therefore the m/z is the M_r.

(b) A fragment with m/z of 17 has been lost. This suggests —OH.

74 Nearly all organic compounds have C—H bonds. These absorb in this region.

75

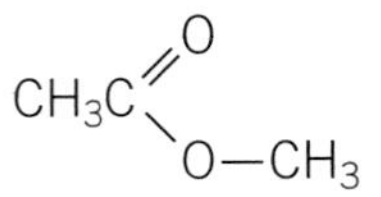

Since this contains —C=O whilst CH_3OCH_3 does not
It is C=O that gives an absorption at 1700 to $1720 \, cm^{-1}$.

76 (a) 2 peaks with δ in range 50 to 70 and in range 5 to 55.

(b) 4 peaks 1 with δ in range 190 to 210 and 3 with δ in range 5 to 55.

77 (a) 1 peak with δ in range 1.9 to 2.9

(b) 3 peaks with peak areas in ratio 3:2:1.
C**H**₃CH₂CHO δ in range 0.7 to 1.6

CH₃C**H**₂CHO δ in range 1.9 to 2.9 CH₃CH₂C**H**O δ in range 9.1 to 10.1

(c) 2 peaks with peak areas in ratio 9:1. δ for both in range 1.0 to 2.0

Answers to exam practice questions

Unit 1

1.1

1 $Ca_3(PO_4)_2$

2 $M(OH)_2$

3 $+6$

4 $Fe_3O_4 + \mathbf{4}CO \longrightarrow \mathbf{3}Fe + \mathbf{4}CO_2$

5 $C_2H_5OH + \mathbf{3}O_2 \longrightarrow \mathbf{2}CO_2 + \mathbf{3}H_2O$

6 $Ca + 2H_2O \longrightarrow Ca(OH)_2 + H_2$

7 $PCl_5 + 4H_2O \longrightarrow H_3PO_4 + 5HCl$

8 $2H^+(aq) + Mg(s) \longrightarrow Mg^{2+}(aq) + H_2(g)$

9 $CH_4 + 4CuO \longrightarrow 4Cu + CO_2 + 2H_2O$

10 $Mg_3N_2 + \mathbf{6}H_2O \longrightarrow \mathbf{3}Mg(OH)_2 + \mathbf{2}NH_3$

11 (a) $\mathbf{4}NH_3(g) + \mathbf{5}\,O_2(g) \longrightarrow \mathbf{4}NO(g) + \mathbf{6}H_2O(g)$

(b)

Element	Initial oxidation number	Final oxidation number
Nitrogen	−3	+2
Hydrogen	+1	+1
Oxygen	0	−2

12 $Ca^{2+}(aq) + CO_3^{2-}(aq) \longrightarrow CaCO_3(s)$

1.2

1 (a)

Particle	Relative mass	Relative charge
proton	1	+1
neutron	1	0

(b) 9 protons, 10 electrons

(c) (i) Number of protons in the nucleus of an atom, e.g. Cl has 17 protons.

(ii) Atoms that have the same number of protons but different numbers of neutrons / same atomic number but different mass number, e.g. Cl-35 has 17 protons and 18 neutrons while Cl-37 has 17 protons and 20 neutrons.

(d)

(e) $1s^2\,2s^2\,2p^6\,3s^2\,3p^6\,4s^1$

2 (a) (i) Time taken for half of the atoms in a sample of a radioisotope to decay (or similar)

(ii) $^{14}_{7}N$

(iii) A neutron changes into a proton (and an electron is emitted by the nucleus)

(iv) 17 190 years

(b) ^{63}Ni

Must be a β-emitter as γ-rays pass easily through thin foil (1)

Must have a long half-life (1)

(no mark for selection without reasoning)

3 (a) (i) 12 protons

(ii) 14 neutrons

(b) (i) 0.125 g

(c) e.g. Cobalt-60 (1) in radiotherapy (1) / Carbon-14 (1) in radio carbon dating (1) / Iodine-131 (1) as a tracer in thyroid glands (1)

(d) (i) ^{218}Po

(ii) Since radon is a gas, α particles will be given off in the lungs (which may cause cancer)

4 (a) (i) $^{40}_{19}K \longrightarrow {}^{40}_{20}Ca + {}^{\,0}_{-1}\beta$ (accept $^{\,0}_{-1}e$)

(1 mark for Ca, 1 mark for balanced equation)

(ii) 3.75×10^9 years

(b) Radioactivity causes mutations / destroys or damages DNA (1)

Alpha radiation causes most damaging / most ionising (1)

Platinum has a long half-life so it emits radioactivity very slowly / bismuth emits radioactivity much more quickly (1)

^{190}Bi is the most damaging (1)

5 (a) C   B   D   E   A

(One mark if one mistake, e.g. C and B wrong way round or A in wrong place)

(b) D (1)

Si is in group 4 therefore large jump in I.E. would be after the fourth I.E. not before it / A, B and C have a large jump before the fourth I.E. so cannot be in group 4 (1)

6 (a) It has greater nuclear charge (1) but little/no extra shielding (1)

(b) Sodium's outer electron is closer to the nucleus (so stronger attraction to the nucleus) (1) and there is less shielding from the inner electrons (1)

(c) (i) Too much / a lot of energy required to form B^{3+} ion

(ii) $K^+(g) \longrightarrow K^{2+}(g) + e^-$

(iii) Value of 1st and 3rd I.E. will be higher (1)

Value of 2nd I.E. will be smaller (1)

(Accept large jump in I.E. value would be between 2nd and 3rd electrons for 1 mark)

7 (a) Ca electronic structure is 2.8.8.2 and Mg is 2.8.2 / Ca has extra shell of electrons (1)

Electrons removed are shielded more / further from nucleus (so are less attracted to the nucleus). (1)

(b) (Electronic structure of Na is 2.8.1.) therefore 2nd electron is removed from inner shell. (1)

Shielding effect is less / attraction to nucleus is greater / closer to nucleus. (1)

8 (a) (i) Only changes between energy levels allowed / electron falls from higher energy levels to lower energy levels (1)

Energy emitted related to frequency / $E = hf$ / the difference between any two energy levels are fixed / energy levels are quantised (1)

(ii)

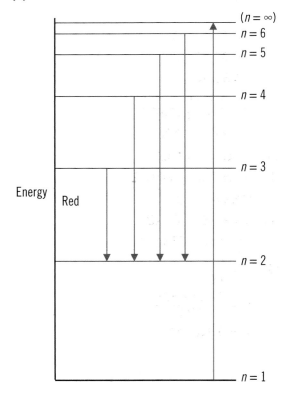

Labelling of any 3 horizontal lines (1)

Transitions going to $n = 2$ (1)

Red line from $n = 3$ to $n = 2$ (1)

(If all lines go to $n = 1$ accept red line from $n = 2$ to $n = 1$)

(iii) Transition from $n = 1$ to $n = \infty$

(b) (i) Hydrogen since frequency is inversely proportional to wavelength

(ii) Hydrogen since energy is proportional to frequency

(c) (i) $f = c/\lambda = 3.28 \times 10^{15}$ (1)

$E = hf$ (1)

2.17×10^{-18} (1)

(ii) Energy = Avogadro's number × energy in (i) (1)

= $6.02 \times 10^{23} \times 2.17 \times 10^{-18}$

= $1\,306\,340\,J = 1306\,kJ\,mol^{-1}$ (1)

9 (a) $1s^2\,2s^2\,2p^3$ (1)

Electrons within atoms occupy fixed energy levels or shells of increasing energy (1)

Electrons occupy atomic orbitals within these shells / The first shell in nitrogen has s orbitals and the second shell s and p orbitals (1)

A maximum of two electrons can occupy any orbital / Each s orbital in nitrogen contains two electrons (1)

Each with opposite spins (1)

Orbitals of the same type are grouped together as a sub-shell / There are three p orbitals in nitrogen's p sub-shell (1)

Each orbital in a sub-shell will fill with one electron before pairing starts / In nitrogen's p sub-shell each orbital contains one electron (1)

(Configuration mark + any 3 from above)

(b) Indicative content

- Atomic spectrum of hydrogen is a series of lines
- Lines get closer together
- As their frequency increases

(Can get these points from labelled diagram)

- Lines arise from atom being excited by absorbing energy
- Electron jumping up to a higher energy level
- Falling back down and emitting energy (in the form of electromagnetic radiation)
- To the $n = 2$ level
- Since lines are discrete, energy levels must have fixed values / Since energy emitted is equal to the difference between two energy levels, ΔE is a fixed quantity or quantum

5–6 marks:
Describes and explains main features fully and states how evidence for energy levels is provided.

The candidate constructs a relevant, coherent and logically structured account including all key elements of the indicative content. A sustained and substantiated line of reasoning is evident and scientific conventions and vocabulary are used accurately throughout.

3–4 marks
Describes main features and gives simple explanation of how features arise.

The candidate constructs a coherent account including most of the key elements of the indicative content. Some reasoning is evident in the linking of key points and use of scientific conventions and vocabulary is generally sound.

1–2 marks
Describes main features only or gives simple explanation of how features arise.

The candidate attempts to link at least two relevant points from the indicative content. Coherence is limited by omission and/or inclusion of irrelevant material. There is some evidence of appropriate use of scientific conventions and vocabulary.

0 marks
The candidate does not make any attempt or give an answer worthy of credit.

1.3

1 27 g

2 **(a)** 172.24 g mol^{-1}

   (b) 20.9%

3 **(a)** **(i)** 1/12th mass of one atom of carbon-12

   (ii) $A_r = \dfrac{(39 \times 93.26) + (40 \times 0.012) + (41 \times 6.73)}{100}$ (1)

   $= 39.14$ (1)

   (Accept 39.13 if 100.002 used)

   (b) To ensure that the ions are not slowed down by air particles

4 **(a)** Mass number 2 is $^2H^+$, Mass number 18 is $(^1H_2O)^+$, Mass number 20 is $(^2H_2O)^+$

   (1 mark if two out of three correct)

   (b) **(i)** Peak at *m/z* 127 (1)

   Peak at *m/z* 254 (1) ignore peak heights

   (ii) Must contain an isotope with a higher relative mass than the stable ^{127}I isotope

5 **(a)** Chlorine-35 and chlorine-37

   (b) 350 molecular ion = four Cl-35

   352 molecular ion = three Cl-35, one Cl-37

   354 molecular ion = two Cl-35, two Cl-37

   356 molecular ion = one Cl-35, three Cl-37

   358 molecular ion = four Cl-37

   (1 mark if three correct)

   (c) Gaseous atoms bombarded by electrons / electron gun to form ions (1)

   Ions accelerated by electric field to high speed (1)

   Deflected through a magnetic field / electromagnet (according to the m/e ratio) (1)

6 **(a)**

Na	Cl	O	
$\dfrac{21.6}{23}$	$\dfrac{33.3}{35.5}$	$\dfrac{45.1}{16}$	(1)
0.939	0.938	2.82	
1	1	3	

   Formula = $NaClO_3$ (1)

   (b) M_r / number of atoms of any element in compound

7 **(a)** Moles $Na_2CO_3 = 0.025 \times 0.045 = 1.125 \times 10^{-3}$ (1)

   Moles $HNO_3 = 2 \times 1.125 \times 10^{-3} = 2.25 \times 10^{-3}$ (1)

   Concentration $HNO_3 = \dfrac{2.25 \times 10^{-3}}{0.0236}$

   $= 0.0953 \, mol \, dm^{-3}$ (1)

   (b) $PV = nRT$

   $n = \dfrac{PV}{RT}$ (1)

   $P = 101\,000 \, Pa$

   $V = 0.0007 \, m^3$

   $T = 301 \, K$ (1)

   $n = \dfrac{101\,000 \times 0.0007}{8.31 \times 301}$

   $n = 2.83 \times 10^{-2} \, mol$ (1)

8 **(a)** Moles HCl = 0.024 (1)

   (b) Moles $CaCO_3$ = 0.012 (1)

   Mass = 1.20 g (1)

   (c) Moles CO_2 = 0.012 (1)

   Volume = 0.288 dm^3 (1)

   (d) $\dfrac{V_1}{T_1} = \dfrac{V_2}{T_2}$

   $V_2 = \dfrac{0.288 \times 323}{298}$ (1)

   $V_2 = 0.312 \, dm^3$ (1)

9 (a) Atom economy $= \dfrac{\text{mass of required product}}{\text{total mass of reactants}} \times 100$ (1)

$= \dfrac{84.01}{58.5 + 17.03 + 44.0 + 18.02} \times 100$

$= 61.1\%$ (1)

(b) Moles NaCl $= \dfrac{900}{58.5} = 15.38$ (1)

Moles $Na_2CO_3 = 7.69$ (1)

Mass $Na_2CO_3 = 7.69 \times 106 = 815(.4)\,g$ (1)

10 Moles MgSO4 $= \dfrac{3.60}{120.4} = 0.030$ (1)

Moles $H_2O = \dfrac{3.78}{18.02} = 0.210$ (1)

$x = 7$ (1)

11 (a) (i) Moles HCl $= \dfrac{0.1 \times 23.15}{1000} = 2.315 \times 10^{-3}$ (1)

Moles $Na_2CO_3 = 1.158 \times 10^{-3}$ (1)

(ii) Moles in original solution $= 1.158 \times 10^{-2}$ (1)

Mass $Na_2CO_3 = 1.227\,g$ (1)

% $Na_2CO_3 = \dfrac{1.227}{2.05} = 59.9\%$ (1)

(b) (i) Only mass of solid needed / all carbonate precipitated out of solution

(ii) Moles $BaCO_3 = \dfrac{2.3}{197} = 1.17 \times 10^{-2}$ (1)

Moles $Na_2CO_3 = 1.17 \times 10^{-2}$

Mass $Na_2CO_3 = 1.17 \times 10^{-2} \times 106 = 1.24\,g$ (1)

% $Na_2CO_3 = \dfrac{1.24}{2.1} \times 100 = 59\%$ (1)

(c) (i) Titration gives more accurate value as it is a mean value (calculated from concurrent results) / uses more accurate or more precise apparatus or technique

(ii) Repeat precipitation / wash precipitate / heat to constant mass /use a more precise balance

(Any two for (1) each)

12 (a) Moles $= \dfrac{0.730}{36.5} = 0.0200$ (1)

Concentration $= \dfrac{0.02}{0.1} = 0.200\,mol\,dm^{-3}$ (1)

Moles in titration $= 0.2 \times 0.0238 = 0.00476$ (1)

(b) Moles MOH in $25.0\,cm^3$ of solution $= 0.00476$ (1)

Moles MOH in original solution $= 0.00476 \times 10$

$= 0.0476$ (1)

(c) $M_r = \dfrac{1.14}{0.0476} = 23.95$ (1)

$A_r = 23.95 - 17.01 = 6.94$

Metal is lithium (1)

13 Moles $SO_2 = \dfrac{1000}{64.1} = 15.6$ (1)

(Ratio of SO_2:SO_3 is 2:2) moles $SO_3 = 15.6$

Theoretical yield of $SO_3 = 15.6 \times 80.1 = 1250\,g$ (1)

% yield $= \dfrac{1225}{1250} \times 100 = 98.0\%$ (1)

14 (a) Moles NaOH = moles unreacted HCl

$= 0.0248 \times 0.188 = 4.66 \times 10^{-3}$ (1)

Moles HCl $= 0.025 \times 0.515 = 1.29 \times 10^{-2}$ (1)

Moles HCl used up in reaction

$= 1.29 \times 10^{-2} - 4.66 \times 10^{-3} = 8.24 \times 10^{-3}$ (1)

(b) Moles $CaCO_3 = \dfrac{8.24 \times 10^{-3}}{2} = 4.12 \times 10^{-3}$ (1)

Mass $CaCO_3 = 4.12 \times 10^{-3} \times 100.1 = 0.412\,g$ (1)

% $CaCO_3 = \dfrac{0.412}{0.497} \times 100 = 82.9\%$ (1)

1.4

1 $Ca^{\times}_{\times} + 2\,^{o}_{o}\overset{oo}{\underset{oo}{Cl}}{}^{o} \longrightarrow Ca^{2+} + 2\left[\,^{o}_{o}\overset{oo}{\underset{oo}{Cl}}{}^{\times}\,\right]^{-}$ (2)

2 A covalent bond is a bond formed by an electron pair between two atoms in which each atom has donated one electron to the bond. In a coordinate bond both electrons in the bond have come from one of the atoms. (2)

3 (a) $\delta- \quad \delta+ \quad \delta- \quad \delta+$

N—H $\quad$ O—Cl (1)

(b) The difference between the electronegativities of Al and Cl of 1.4 units is less than that between Al and O of 1.9 units so that the electrons are more equally shared in the chloride and the bonds are more covalent in nature. (1)

4 (a) $\delta+ \quad \delta-$

Cl—F (1)

(b) $^{o}_{o}\overset{oo}{Cl}\,^{\times}_{o}\overset{\times\times}{F}{}^{\times}_{\times}$ (1)

5 (a) Iodine – van der Waals force, Diamond – covalent bond (2)

(b) Diamond since the covalent bond strength is much greater than the van der Waals force. (1)

6 van der Waals forces < hydrogen bonds < covalent bonds (1)

7 (a) There are three electrons in BF_3 forming bonding pairs with fluorine. The identical bonds are equally arranged in a plane giving bond angles of 120°. In NH_3 there are again three bond pairs but also a lone pair on the nitrogen.

The increased lone pair – bond pair repulsion given by VSEPR distorts the three bond pairs out of the nitrogen plane to give a trigonal pyramid. The orbital shape is tetrahedral although there is no bond in the lone pair position. However, since the lone pair – bond pair repulsion is greater than bond pair – bond pair, the bond angle in NH_3 is 107°, less than the tetrahedral value of 109.5°. (2)

(b) (i) A coordinate bond (1)

(ii) Value of 109.5°. The nitrogen lone pair forming the bond will provide lone pair – bond pair repulsion in the BF_3 and give a tetrahedral shape. (1)

8 (i) Methane has four equal bond pairs and no lone pairs so that the symmetrical tetrahedral shape is formed. (2)

(ii) In water the orbital shape is essentially tetrahedral but the two lone pairs on the oxygen repel the two bond pairs more strongly than they repel one another so that the tetrahedral bond angle of 109.5° is reduced to 104.5°. (2)

9 (i) Aluminium and boron are in Group 3 and thus have only three bonding electrons and no lone pairs. They therefore form three identical bond pairs with chlorine with equal repulsion to give trigonal planar molecules having bond angles of 120°. Nitrogen, however, is in Group 5 and has a lone pair remaining after its three bond pairs with chlorine atoms. (3)

(ii) Using VSEPR it is seen that the greater lone pair – bond pair repulsion will distort the otherwise planar trigonal arrangement into a tetrahedral shape with a 109.5° bond angle. (3)

10 SF_6, H_2O (2)

11 (a) In SO_2 sulfur is combined with two oxygens in a neutral molecule and therefore has an oxidation number of 4. It reacts with two fluorine atoms that pass from oxidation numbers of zero to −1, that is they are reduced while oxidising the sulfur atom from 4 to 6. The answer may also be obtained by inspection of the neutral molecule with the sulfur balancing out two oxygens and two fluorine atoms. (2)

(b) The molecule has no lone pairs but four bonds, two double bonds to oxygen atoms and two single bonds to fluorine atoms. The VSEPR theory treats single and double bonds as the same so that the four bonds from the sulfur atom form a tetrahedron. (1)

12 Boiling temperatures (b.t.) generally increase with molecular mass since the number of electrons involved in induced dipole interactions also increases. Thus the b.t. of HCl < HBr < HI. However, a different intermolecular force exists in the case of HF in

which the small, electronegative fluorine atom forms hydrogen bonds that are stronger than the induced dipole forces. Thus the b.t. of HF that might have been expected to be about 160 K is over 100° higher. (4)

1.5

1 All ○ are Br^- and ● are Li^+.

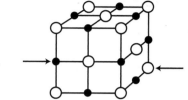

(1)

2 (a) (i)

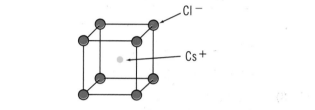

(2)

(ii) The Na ions are smaller than the Cs ions so that only six of the larger chloride anions are able to pack around them as against eight around the larger Cs ion. (1)

(b)

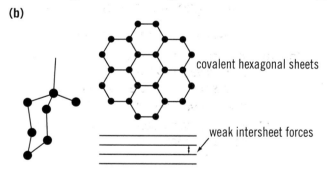

covalent hexagonal sheets

weak intersheet forces

(i) In diamond four strong covalent bonds connect each carbon atom to four others in a giant tetrahedral structure.

Graphite forms a planar hexagonal layer structure in which each carbon atom is bonded to three others in a delocalised system. The planes are held together in the crystal by weak intermolecular forces. (2)

(ii) Both diamond and graphite have high melting temperatures owing to the giant molecular strong covalent bonding but graphite is soft owing to the weak intermolecular forces while diamond is extremely hard. (2)

3 The covalent bonds in iodine are used only to bond two iodine atoms to form the I_2 molecules that are only held together in the simple molecular crystal by weak van der Waals forces. The strong covalent bonds in diamond need an energy corresponding to a temperature of over 3000 degrees to be broken while the weak intermolecular forces holding the I_2 molecules together in the crystal will break at around 100 degrees. (3)

4 It contains delocalised electrons in an electron sea that are free to carry the electric current.

5 **(i)** Diamond is a giant covalent structure with strong tetrahedral bonding. Metals have a different strong bonding in which each metal atom donates one or more electrons to a delocalised electron 'gas' or 'sea' that bonds the positive ions formed by the donation through the attraction between opposite charges. The positive ions are commonly close-packed. (3)

(ii) Both diamond and metals are strongly bonded and have high melting temperatures (usually so in metals). Metals are good electrical conductors through their delocalised electrons while diamond is an insulator since all the electrons are tied up in the four covalent bonds. (3)

6 **(a)** $Fe_2O_3 + \mathbf{3}CO = \mathbf{2}Fe + \mathbf{3}CO_2$. (1)

(b) Iron is in oxidation state 3 in Fe_2O_3 and is reduced to oxidation state zero for the element by the CO. The carbon is CO is in oxidation state 2 and is oxidised to state 4 as it reduces the iron oxide. (2)

(c) **(i)** 6:6 (1)

(ii) The drawing should show a symmetrical arrangement of six oxide anions around the smaller Fe(II) cation in the same way that six chloride anions are arranged around the sodium cation. (1)

(d) In covalent bonds both bonding atoms contribute one electron to the bond pair so that each atom contributes two electrons in CO. The coordinate bond is formed by one atom (in this case the oxygen) contributing both electrons to the bond. (2)

(e) In metallic bonding each atom donates one or more electrons to a delocalised electron 'sea' or 'gas' leaving a positive ion lattice that is held together by electrostatic attractions between the cations and the electron sea. The bonding is very strong so that iron does not melt until a temperature of above 1500°C is reached. (4)

1.6

1 **(a)** **(i)** $Ba^{2+} + SO_4^{2-} \longrightarrow BaSO_4$ (1)

(ii) A white precipitate of barium sulfate is seen. (1)

(b) **(i)** Apple green (1)

(ii) Adding silver nitrate solution give a white precipitate of silver chloride. (2)

2 Only one outer electron is involved in metallic bonding in sodium as against two in magnesium. The stronger bonding means that a higher temperature is required to melt the magnesium. (3)

3 **(i)** Ne, Ar, Cl. (1)

(ii) Both silicon and carbon (in diamond) contribute four electrons to form four strong covalent bonds that are symmetrically arranged to give very stable structures that require the energy of high temperatures to give melting. (2)

(iii) Neon and argon have full electron shells and form no bonds with one another so that only weak induced dipole bonds are present to hold the atoms together in a condensed phase. (2)

(iv) They have one, two and three electrons respectively available for metallic bonding so that the binding energy increases and higher temperatures are needed to melt the solids. (2)

(v) Carbon is a smaller atom than silicon and a more compact structure having stronger covalent bonding can be formed. (1)

(vi) Argon has a greater atomic mass than neon with more electrons available for induced dipole interatomic forces so that a higher temperature is needed to separate the atoms. (1)

4 Five of the seven electrons in Group 7 are p-electrons and the elements have the characteristic electronic structures and chemical properties of the p-block. (1)

5 **(i)** They all tend to gain an electron and form a stable negative ion, X^-, having a full outer electronic shell. This is the process of oxidation, that is the species that loses this electron is oxidised. (1)

(ii) Fluorine is the strongest oxidising agent, the most electronegative element and will oxidise all the other halides to form a fluoride and the free element. (1)

(iii) Five (1)

6 **(i)** Brick red (1)

(ii) A cream-white precipitate was formed. (1)

(iii) $Ag^+ + Br^- = AgBr$ (1)

(iv) A reddish-brown colour was seen corresponding to the liberation of bromine by the oxidising chlorine. (1)

(v) Chlorine is a stronger oxidising agent than bromine and therefore oxidises the bromide ion to bromine while being reduced to chloride. The bromide anion was present as an ionic compound with calcium ions.

The oxidising equation is: $Br^- + \frac{1}{2}Cl_2 = \frac{1}{2}Br_2 + Cl^-$. (2)

7 Barium carbonate (1)

$Ba^{2+} + CO_3^{2-} \longrightarrow BaCO_3$ (1)

8 $MgCO_3.Mg(OH)_2.3H_2O \longrightarrow 2MgO + CO_2 + 4H_2O$

MgO as product [1] correct balancing [1]

9 (i)

	K_2CO_3	NaOH	$BaCl_2$	$MgCl_2$
K_2CO_3	–	nvr	ppt	ppt
NaOH	nvr	–	nvr	ppt
$BaCl_2$	ppt	nvr	–	nvr
$MgCl_2$	ppt	ppt	nvr	–

Nvr is no visible reaction; all ppts are white. (4)

(ii) Magnesium carbonate;

$Mg^{2+} + CO_3^{2-} \longrightarrow MgCO_3$ (2)

(iii) Potassium carbonate and sodium hydroxide turn litmus red.

Potassium lilac flame, sodium yellow, barium apple green, magnesium white (or no colour).

Only barium reacts giving a white ppt of barium sulfate. (4)

1.7

1 (a) (i) $2NH_3 + H_2SO_4 \longrightarrow (NH_4)_2SO_4$

(ii) It accepts a proton / H^+

(b) (i) Fewer gas particles in product (1)

Equilibrium shifts towards product to reduce pressure (1)

(ii) Equilibrium shifts towards product (1)

to replace ammonia (that is removed) (1)

(c) No change

(d) Rate of reverse endothermic reaction increases (1)

Equilibrium shifts towards reactants therefore K_c decreases (1)

2 (a) (i) Increased CO_2 pushes the equilibrium to the right. (1)

pH will fall since $[H^+]$ increases (1)

(ii) This will decrease since the increase in H^+ moves equilibrium to the right, (reducing carbonate and increasing hydrogencarbonate)

(b) (i) A reversible reaction where the forward and reverse reactions occur at the same rate

(ii) (Molecules can escape from the bottle) so concentration /amount / pressure of $CO_2(g)$ falls (1) and position of equilibrium moves to the left (so concentration of $CO_2(aq)$ falls) (1) rate of molecules entering solution is less than rate leaving solution. (1)

(Accept any two points)

3 (a) (i) $K_c = \dfrac{[CO][H_2]^3}{[CH_4][H_2O]} mol^2 dm^{-6}$

expression (1), units (1)

(ii) If a system at equilibrium is subjected to a change then the position of equilibrium will shift to minimise that change.

(iii) I Yield of hydrogen increases (1)

Equilibrium moves in forward endothermic direction (1)

II Yield of hydrogen decreases (1)

Equilibrium moves towards side with smaller number of gas molecules (1)

(b) (i) I Yield decreases

II Yield increases

(ii) Equilibrium moves in backward endothermic direction (1)

Equilibrium moves towards side with smaller number of gas molecules (1)

4

Change	Effect, if any, on position of equilibrium	Effect, if any, on value of K_c
Addition of reactant at constant temperature	Shift to the right	None
Drop in temperature	Shift to the right	Decreases

5 (a) (i) Acidic solutions have a pH of less than 7 (1)

The lower the figure the greater the degree of acidity (1)

(ii) $pH = -\log[H^+]$, $[H^+] = $ antilog -5.5 (1)

$[H^+] = 3.2 \times 10^{-6} mol dm^{-3}$ (1)

(b) (i) It produces $H^+(aq)$ ions on reaction with water

(ii) Concentration of hydrogen ions would increase (1)

as position of equilibrium moves to the right (1)

(c) A weak acid is one that partially dissociates in aqueous solution (1)

A dilute acid is one where a small amount of acid has been dissolved in a large volume of water (1)

6 **(a)** Weighing bottle would not have been washed / difficult to dissolve solid in volumetric flask / final volume would not necessarily be 250 cm³

(b) Pipette

(c) To show the end point / when to stop adding acid

(d) So that a certain volume of acid can be added quickly before adding drop by drop / to save time before doing accurate titrations / to give a rough idea of the end point

(e) To obtain a more reliable value

7 **(a)** Moles CuSO$_4$.5H$_2$O = 0.25 × 0.25 = 0.0625 (1)

Mass CuSO$_4$.5H$_2$O = 0.0625 × 249.7 = 15.6 g (1)

(b) Indicative content

- Transfer solid into a beaker and add water
- Stir until all the solid dissolves
- Wash out original vessel
- Add solution to a 250 cm³ volumetric flask
- Use of a funnel
- Make up to the mark (with distilled water)
- Shake / invert

5–6 marks:

Describes, giving full practical details, how solution is prepared.

The candidate constructs a relevant, coherent and logically structured account including all key elements of the indicative content. A sustained and substantiated line of reasoning is evident and scientific conventions and vocabulary are used accurately throughout.

3–4 marks

Describes, giving main practical details, how solution is prepared.

The candidate constructs a coherent account including most of the key elements of the indicative content. Some reasoning is evident in the linking of key points and use of scientific conventions and vocabulary is generally sound.

1–2 marks

Describes some details of how solution is prepared.

The candidate attempts to link at least two relevant points from the indicative content. Coherence is limited by omission and/ or inclusion of irrelevant material. There is some evidence of appropriate use of scientific conventions and vocabulary.

0 marks

The candidate does not make any attempt or give an answer worthy of credit.

8 **(a)** Identification of 23.95 cm³ as anomalous result (1)

Mean titre = 23.25 cm³ (1)

(b) 25.00 cm³ of the potassium carbonate solution **pipetted** into a conical flask (1)

(A few drops of) indicator added (1)

Titrate (with the acid) until the indicator just (1) turns pink (1)

Shake/swirl/mix (1)

Reads burette before and after (1)

(Need first point + any other four points)

(c) Funnel left in burette (1), air in pipette (1), not reading meniscus (1), solution in flask not mixed thoroughly (1), all of solid not used to make solution (1)

(Maximum 2 marks for sources of error)

If end-point overshot, too much acid would have been added (1), so moles (mass) carbonate calculated would have been more than actual moles (mass) present (1)

(d) More accurate result using 0.2 mol dm⁻³ HCl since it requires significant volume (1)

Percentage error is greater in measuring smaller volume (1)

Unit 2

2.1

1 **(a)** $\Delta H = (2 \times -394) + (3 \times -286) - (-1560)$ (1)

$\Delta H = -86\,kJ\,mol^{-1}$ (1)

(b) The enthalpy change when one mole of a substance is formed from its constituent elements (1) in their standard states under standard conditions (1)

(c) **(i)** $H_2(g) + \frac{1}{2}O_2(g) \longrightarrow H_2O(g)$

(ii) $-242 = 436 + 248 - 2(O-H)$ (1)

$2(O-H) = 926$

$O-H = 463\,kJ\,mol^{-1}$ (1)

2 **(a)** ΔH reaction $= \Delta_f H$ products $- \Delta_f H$ reactants (1)

$-46 = \Delta_f H$ ethanol $- (52.3 - 242)$

$\Delta_f H$ ethanol $= -46 - 189.7$ (1)

$\Delta_f H$ ethanol $= -235.7\,kJ\,mol^{-1}$ (1)

(b) Bonds broken $= 1648 + 612 + 926$

$= 3186\,kJ\,mol^{-1}$ (1)

Bonds formed $= 2060 + 348 + 360 + 463$

$= 3231\,kJ\,mol^{-1}$ (1)

ΔH reaction $= 3186 - 3231 = -45\,kJ\,mol^{-1}$ (1)

(c) **(i)** Average bond enthalpies used not actual ones

(ii) Yes, since answers are close to each other

(d) **(i)** $C_2H_5OH + 3O_2 \longrightarrow 2CO_2 + 3H_2O$

(ii) $\Delta H = (2 \times -394) + (3 \times -286) - (-278)$ (1)

$\Delta H = -1368\,kJ\,mol^{-1}$ (1)

(iii) Energy for ethanol $= \dfrac{1368}{46} = 29.7\,kJ\,g^{-1}$ (1)

Energy for octane $= \dfrac{5512}{114} = 48.4\,kJ\,g^{-1}$ (1)

(iv) Ethanol is a renewable fuel (if obtained by fermentation) / Ethanol is cheaper in countries with a plentiful sugar cane growth / Ethanol is more carbon neutral / Ethanol burns more cleanly

3 **(a)** To ensure that the (initial) temperature is constant / temperature difference is required between initial and maximum temperature.

(b) Heat $= 50 \times 4.18 \times 9.6$ (1)

Heat $= 2006\,J$ (1)

(c) Moles $CuSO_4 = 0.025$ (1)

$\Delta H = -\dfrac{2006}{0.025}$ (1)

$= -80.2\,kJ\,mol^{-1}$ (1)

(d) Burette / pipette

(e) Magnesium was in excess.

(f) Larger surface area (1)

Rate of reaction is quicker (1)

(g) $\dfrac{12.9}{93.1} \times 100 = 13.9\%$

(h) Heat / energy is lost to the environment. (1)

States how insulation is improved, e.g. place a lid on the polystyrene cup. (1)

4 **(a)** **(i)** Moles acid $= 0.5 \times 0.05 = 0.025$

(ii) Heat $= 100 \times 4.18 \times 3.4$ (1)

Heat $= 1421\,J$ (1)

(iii) $\Delta H = -\dfrac{1421}{0.025}$ (1)

$= -56.8\,kJ\,mol^{-1}$ (1)

(b) **(i)** The total enthalpy change for a reaction is independent of the route taken from the reactants to the products.

(ii) If a reaction was not independent of the route, it would be possible to create energy by making the products via the intermediate by one route and then converting back to the reactants by the other route. This would be contrary to the law of conservation of energy.

(c) $\Delta H = (4 \times -176) - (2 \times -242)$ (1)

$\Delta H = -220\,kJ\,mol^{-1}$ (1)

5 **(a)** Bonds broken $= (1\ C{=}C) + (4\ C-H) + (1\ H-H)$

Bonds formed $= (6 \times C-H) + (1\ C-C)$ (1)

$-124 = (C{=}C) + (4 \times 412) + 436 - ((6 \times 412) + 348)$ (1)

$(C{=}C) = -124 - 2084 + 2820 = 612\,kJ\,mol^{-1}$ (1)

(b) $\Delta H = 39\,kJ\,mol^{-1}$

(c) **(i)** $\Delta H = (2 \times -286) + (2 \times -297) - (2 \times -20.2)$ (1)

$\Delta H = -1125.6\,kJ\,mol^{-1}$ (1)

(ii) Oxygen gas is an element in its standard state

6 **(a)** $M_r(C_9H_{20}) = 128$ (1)

Moles $= 1.563 \times 10^{-3}$ (1)

(b) Temperature increase $= 30.7\,°C$ (1)

Heat $= 50 \times 4.18 \times 30.7 = 6416\,J$ (1)

$\Delta H = -4105\,kJ\,mol^{-1}$ (1)

(c) Heat loss to the environment / incomplete combustion (1)

Lag calorimeter / use a lid (1)

2.2

1 (a) Measure (the volume of) hydrogen produced (using a gas syringe) / (mass of) hydrogen lost at constant time intervals

(b) Crush it into a powder / increase its surface area / heat it / stir it

2 (a) Peak of curve must be lower and to the right of the original curve

(b) (i) Rate = $\dfrac{0.0006}{200}$ = $3.0 \times 10^{-6}\,\text{mol}\,\text{dm}^{-3}\,\text{s}^{-1}$

(ii) It would decrease   (1)

Since the concentration of but-1,3-diene decreases   (1)

(iii)

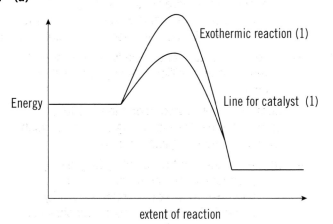

3 (a) Rate increases   (1)

More molecules will have required activation energy   (1)

(b) Rate increases   (1)

More molecules in given volume, greater chance for effective collisions   (1)

4 (a)

(b) If there is an increase in temperature there will be an increase in kinetic energy of the molecules (1). More molecules will have an energy greater than the activation energy (1) so a greater fraction of collisions are effective, meaning that the reaction rate will increase (1)

5 (a) The diagram shows two reasonable distribution curves with T_2 flatter and 'more to the right' than T_1. (1)

Activation energy correctly labelled, or mentioned in the writing   (1)

Fraction of molecules having the required activation energy is much greater at a higher temperature (increasing the frequency of successful collisions) (in words)   (1)

(b) Place the mixture on a balance and measure the (loss in) mass (1) at appropriate time intervals (1)

Or other suitable method, e.g. sample at intervals / quench (1) titration (1)

6 Lower temperatures can be used   (1)

Energy costs saved   (1)

More product can be made in a given time (so more can be sold)   (1)

Enable reactions to take place that would be impossible otherwise   (1)

Less (fossil) fuels burned to provide energy (so less CO_2 formed)   (1)

(Accept any 3 points)

7 (a) Bubbles (of gas) / fizzing / (some) $CaCO_3$ disappears / apparatus gets warmer.

(b) Gas syringe / (inverted) burette.

(c) (Use scales to) weigh aqueous product / sampling and titration / measure change in pH at set times.

(d) (i) Smooth curve passing through $150\,\text{cm}^3$ ending at $200\,\text{cm}^3$

(ii) Curve less steep (1) ending at $100\,\text{cm}^3$ (1)

(iii) When the acid is less concentrated it has fewer (acid) particles (1) therefore there is less chance of successful collisions (between the acid and carbonate) / fewer collisions per unit time. (1)

(e) Diagram with two reasonable curves.   (1)

Labelling for activation energy.   (1)

The fraction of molecules that have the required activation energy is much greater at a higher temperature.   (1)

8 (a) Correct plotting – within half a small square (2), one error (1)

Appropriate best fit line drawn (1)

(b) (i) C   (1)

Curve steeper   (1)

(ii) Concentration of acid is greatest

(c) $44\,\text{cm}^3$ ($\pm 1\,\text{cm}^3$)

(d) Lower the temperature of the acid (1)

Reactants collide with less energy (1)

Fewer molecules that have the required activation energy (1)

or Use pieces of magnesium (1) less surface area (1) less chance of successful collisions (1)

9 (a) Temperature (1), pressure / concentration (1), catalyst (1), particle size (1), light (1)

(Accept any three)

(b) (i) Correct plotting – within half a small square (2), one error (1)

Good curve and tangent (1)

Rate 0.10 ± 0.01 (1)

Units $cm^3 s^{-1}$ (1)

(ii) Rate decreases as reaction proceeds (1) since concentration of peroxide decreases (fewer successful collisions) (1)

(iii) Gas syringe attached to reaction flask, (1) stopwatch started and volume of gas measured at set time intervals (1)

(c) Molecules (A and B) must collide with one another (1) and have enough energy to react (to form C) (1)

Increasing pressure means that there are more molecules in a given volume so greater chance for effective collisions (1)

Increasing temperature means that more molecules will have the required (activation) energy to react (1)

10 (a) At least 5 points plotted correctly (1)

Appropriate straight line drawn (1)

Axes correctly labelled (1)

(b) Rate $1.1 \times 10^{-4} \pm 0.1 \times 10^{-4}$ (1)

Units $mol\,dm^{-3}\,s^{-1}$ (1)

(c) Colorimetry method.

Calibrate colorimeter (1) with iodine solution of known concentration (1). Measure light passing through to determine concentration at intervals (1)

(d) Concentration of hydrogen peroxide is directly proportional to the rate / doubling the concentration of hydrogen peroxide doubles the rate (1)

Concentration of iodide ions is directly proportional to the rate / doubling the concentration of iodide ions doubles the rate (1)

2.3

1 (a) $-55.6\,kJ\,g$ $-48.5\,kJ\,g$ (2)

(b) $2.75\,g$ $3.03\,g$ (2)

(c) Methane since butane produces more CO_2 per kg of energy produced

[butane 0.062; methane 0.049] (2)

2 Ethane produces more energy per mol than bioethanol (1560 kJ as against 1371 kJ).

However, ethane is a non-renewable fossil fuel, unlike bioethanol that may be renewed by growing sugar cane. It is thus likely that bioethanol is the more friendly.

Note that both combustions produce the same amount of CO_2/ per mol of fuel so that bioethanol generates slightly more CO_2 in producing the same amount of energy. (3)

3 (a) (i) Reaction **1** absorbs the most CO_2 per mol of carbonate used. (2)

(ii) If a system in equilibrium is subjected to a change, the equilibrium tends to shift in order to minimise the effect of that change. (1)

(iii) Increasing gas pressure will cause the equilibrium to shift to the right so that CO_2 removal will be more efficient. (2)

(b) (i) 6.25 (1)

(ii) $150\,dm^3$ (1)

(iii) 15% (1)

(c) (i) An acid is a proton donor. (1)

(ii) Carbon dioxide dissolves in water to form carbonic acid that weakly dissociates to give hydrogen ions:

$CO_2 + H_2O = H_2CO_3 = H^+ + HCO_3^-$ (1)

(iii) The atmosphere contains 400 ppm of carbon dioxide that dissolve in rainwater to give a weakly acid solution of pH around 5. (1)

4 (a) Crude oil is a non-renewable fossil fuel formed several hundred million years ago and will run out in time.

Combustion of hydrocarbon fuels produces carbon dioxide, the increasing amount of which may be a cause of global warming.

(b) (i) While the combustion of hydrogen only produces water, the manufacture of hydrogen may require much electricity generated by fuel-burning power stations. Also oxides of nitrogen may be generated in combustion engines that can cause pollution and smog. (2)

(ii) It is true that hydrogen/air mixtures may be explosive over certain ranges. Also hydrogen leaks from the system could be dangerous. (1)

(c) A wide variety of answers is acceptable bearing in mind the scope of the topic as listed in a summary below. (5)

1 Prevent waste

2 Increase atom economy

3 Use safer methods, chemicals and solvents

4 Increase energy efficiency

5 Use renewable raw materials (feedstocks)

6 Use catalysts (vs stoichiometric reactions)

7 Prevent pollution and accidents

8 Design for biodegradation.

The question of energy use and its effects with respect to global warming is clearly a major factor.

5 (a) (i) T is 298 K; P is 1 atmosphere or 10^5 Pascals (2)

(ii) The standard enthalpy of formation of an element is the energy change in forming the element from itself under standard conditions, i.e. zero. (1)

(iii) -174 kJ mol (2)

(b) (i) No effect since there is the same number of gaseous molecules on either side of the equilibrium. (2)

(ii) Since the forward reaction is exothermic an increase in temperature will drive the reaction to the left, the endothermic direction, and decrease the yield of hydrogen. (2)

(iii) Catalysts have no effect on the position of equilibrium only on the rate at which equilibrium is reached. (1)

2.4

1 (a) A process of bond breaking where the two electrons go to one of the atoms involved in the bond.

(b) $(CH_3)_3C^+$ and Cl^-

2 (a) Alkene

Alcohol

(b) $C_5H_{10}O$

3 (a) Compounds that have the same molecular formula but different structural formulae.

(b) There is restricted rotation about a double bond.

Both carbon atoms in the double bond have different atoms/groups bonded to them in compound B.

(c) M_r of compound A = 146.3

Cost per mol $= \dfrac{146.3 \times 48 \times 100}{100 \times 73} = £96.19$

4 Hex-2-ene

2.5

1 (a) It contains an unpaired electron.

(b) (i) $^\bullet C_3H_7 + Cl_2 \longrightarrow C_3H_7Cl + Cl^\bullet$

(ii) A radical reacts to produce another radical and the process continues.

(c) C_7H_{16}

(d) A covalent bond breaks and each atom receives one electron.

2 (a) A compound that contains hydrogen and carbon only and has no multiple bonds.

(b) (i) $C_3H_8 + 5O_2 \longrightarrow 3CO_2 + 4H_2O$

(ii)

(c) Ethene has σ bonds between carbon and hydrogen atoms

It also has a π bond between the 2 carbon atoms

This is formed when p electrons (one on each carbon) overlap sideways.

(d) M_r ethanol = 46 and M_r glucose = 180

230 g of ethanol is 5 moles so from equation 2.5 moles of glucose needed

Mass glucose = 2.5 × 180 = 450 g

3 (a) Molecules in different fractions have different numbers of carbon atoms.

More carbon atoms mean larger molecules and therefore greater van der Waals' forces between the molecules.

(b) Mass of petroleum gases = 1.2% × 145 000 = 1740 g

$$\text{Moles of butane} = \frac{1740}{58} = 30$$

Volume of butane = 30 × 24 = 720 dm³

(c) Brent crude would be better as it has more naphtha

Any equation that gives ethene e.g.

$C_{10}H_{22} \longrightarrow C_2H_4 + C_8H_{18}$

Polymerisation: many small molecules joining together to make a large molecule.

This is addition polymerisation

Many possible e.g. poly(phenylethene) (polystyrene), poly(chloroethene) (PVC), poly(tetrafluoroethene) (PTFE) and relevant monomer

2.6

1 **(a)** $C_6H_{12}Br_2$

(b) Elimination

2 **(a)** C—Cl bond in 1,1,1-trichoroethane is weaker than the C—H and the C—C in methylcyclohexane

C—Cl bond broken by UV light to produce radicals that damage the ozone layer

(b) Add bromine

Hept-2-ene decolourises the bromine but methylcyclohexane does not

3 Indicative content

- To produce halide negative ion C–halogen bond needs to be broken

- C–halogen bond becomes weaker as group is descended/ with increase in halogen size

- This would suggest that the time for hydrolysis would become shorter as group descended

- Hydrolysis is nucleophilic attack on δ+ carbon

- Carbon becomes less δ+ as group descended

- This would suggest that the time for hydrolysis would become longer as group descended

- From data table, time is shorter so effect of bond strength outweighs effect of bond polarity.

5–6 marks:

Both bond polarity and bond strength considered. Effect on time of each factor identified. Clear conclusion of nature of dominant factor made.

The candidate constructs a relevant, coherent and logically structured account including all key elements of the indicative content. A sustained and substantiated line of reasoning is evident and scientific conventions and vocabulary are used accurately throughout.

3–4 marks

Both bond polarity and bond strength considered. Effect on time of each factor identified.

The candidate constructs a coherent account including most of the key elements of the indicative content. Some reasoning is evident in the linking of key points and use of scientific conventions and vocabulary is generally sound.

1–2 marks

Both bond polarity and bond strength considered in some way,

The candidate attempts to link at least two relevant points from the indicative content. Coherence is limited by omission and/ or inclusion of irrelevant material. There is some evidence of appropriate use of scientific conventions and vocabulary.

0 marks

The candidate does not make any attempt or give an answer worthy of credit.

2.7

1 **(a)** **(i)** Aqueous sodium hydroxide

Heat

(ii) To dissolve in water hydrogen bonds are needed

Butan-1-ol contains OH

This means that it forms hydrogen bonds with water.

(b) **(i)** Acidified potassium dichromate/ manganate(VII) and heat

(ii) Fractional distillation.

2 **(a)** **(i)** δ+ on H and δ− on Br

Arrow showing movement of pair of electrons from the H—Br bond onto Br and from C=C to δ + H

Positive ion with + on second C

Arrow from Br⁻ to form bond with middle carbon to give the product

Mechanism is electrophilic addition.

(ii) Atom of Br can be added to either carbon in C=C

Secondary carbonium ions are more stable than primary carbonium ions

(b) IR peak at $1715\,cm^{-1}$ due to C=O

δ 9.8 due to R—CHO

Molecule must be propanal

Arises from compound **C** as aldehydes formed from primary alcohols only

(c) **(i)** Colour change should be orange to colourless

Product should be 1,2-dibromopropane

(ii) Would see white precipitate due to formation of AgCl from use of HCl

Should warm with aqueous sodium hydroxide, then acidify with nitric acid and then add aqueous silver nitrate.

2.8

1 (a) (i)

(ii) 2-Bromopropane formed from a secondary carbocation

Secondary carbocations are more stable than primary carbocations

(b) Indicative content

- Ratio C : H : Br = $\dfrac{29.8}{12} : \dfrac{4.2}{1} : \dfrac{66.0}{80}$

  = 2.48 : 4.20 : 0.825

  = 3 : 5 : 1

  Empirical formula is C_3H_5Br

- In mass spectrum two molecular ion peaks are due to two isotopes of bromine.

- From mass spectrum M_r with ^{81}Br = 122. M_r of C_3H_5Br = 122.

  Molecular formula is C_3H_5Br

- Mass spectrum peak at 15 due to CH_3^+ and at 41 due to $C_3H_5^+$

- In infrared spectrum absorption at 550 cm^{-1} due to C—Br, at 1630 cm^{-1} due to C=C and at 3030 cm^{-1} due to C—H

- Molecule shows *E-Z* isomerism so must have two different groups attached to each carbon in the double bond

- Molecule is

5–6 marks:
Empirical formula determined, molecular formula determined, all points in both spectra considered, significance of *E-Z* noted. Correct final formula.

The candidate constructs a relevant, coherent and logically structured account including all key elements of the indicative content. A sustained and substantiated line of reasoning is evident and scientific conventions and vocabulary are used accurately throughout.

3–4 marks
Empirical formula determined, molecular formula determined, some points in both spectra considered.

The candidate constructs a coherent account including most of the key elements of the indicative content. Some reasoning is evident in the linking of key points and use of scientific conventions and vocabulary is generally sound.

1–2 marks
Minimum of either empirical or molecular formula determined. At least one valid comment on each spectrum given.

The candidate attempts to link at least two relevant points from the indicative content. Coherence is limited by omission and/ or inclusion of irrelevant material. There is some evidence of appropriate use of scientific conventions and vocabulary.

0 marks
The candidate does not make any attempt or give an answer worthy of credit.

2 (a) Any three absorption peaks labelled
O—H at 3200–3500; C=O at about 1750; C—O at 1000–1200; C—H at 3000–3100.

(b) M_r is 60

Fragments at 15 due to CH_3^+, fragment at 45 due to $COOH^+$.

(c) (i) Heat with acidified potassium dichromate(VI) or acidified potassium manganate(VII).

(ii) Colour change orange to green or purple to colourless/pale pink.

(d) (i) Expect propane to have a lower boiling temperature because ethanol can hydrogen bond and propane cannot.

(ii) Expect butan-1-ol to have a higher boiling temperature because it is larger and therefore has more intermolecular/ Van der Waal's forces.

Index